VOCATIONAL EDUCATION AND THE A
LC5225.E57 V63 1992

DATE DUE

FE 17 '95			

DEMCO 38-296

VOCATIONAL EDUCATION AND THE ADULT UNWAGED

NEW DEVELOPMENTS IN VOCATIONAL EDUCATION

Series Editors: Peter Funnell and Dave Muller

VOCATIONAL EDUCATION AND THE ADULT UNWAGED

DEVELOPING A LEARNING CULTURE

JENNY HUNT

HEATHER JACKSON

KOGAN PAGE

Acknowledgements

We would like to thank our families, friends and colleagues for all their help, support and patience, especially the members of Consortium Five Training Consultants.

We would also like to thank Elaine Pole and the staff at NIACE for their help and time in providing source material during the research phase of the book.

Jenny Hunt and Heather Jackson

First published in 1992

Apart from any fair dealing for the purposes of research or private study, or criticism or review, as permitted under the Copyright, Designs and Patents Act, 1988, this publication may only be reproduced, stored or transmitted, in any form or by any means, with the prior permission in writing of the publishers, or in the case of reprographic reproduction in accordance with the terms of licences issued by the Copyright Licensing Agency. Enquiries concerning reproduction outside those terms should be sent to the publishers at the undermentioned address:

Kogan Page Limited
120 Pentonville Road
London N1 9JN

© Jenny Hunt, Heather Jackson and named contributors, 1992

British Library Cataloguing in Publication Data
A CIP record for this book is available from the British Library
ISBN 0 7494 0493 0

Typeset by Paul Stringer, Watford
Printed and bound in Great Britain by
Biddles Ltd, Guildford and King's Lynn

Contents

Series Editors' Foreword	8
Foreword	10
General Introduction	12
List of Contributors	13

Part I   Introductory Perspectives 15

Introduction 15

Chapter 1   The Unwaged and the Unemployed 17
Jenny Hunt and Heather Jackson

Chapter 2   Adult Vocational Education and Training: What, Where and When? 25
Jenny Hunt and Heather Jackson

Chapter 3   Factors Influencing the Need for Change 36
Jenny Hunt and Heather Jackson

Chapter 4   The Training Debate 44
Jenny Hunt and Heather Jackson

Part II   Vocational Education and Training Provision for Unwaged Adults 53

Introduction 53

Chapter 5   Six Key Aspects in the Delivery of Adult Vocational Education and Training 57
Jenny Hunt and Heather Jackson

Chapter 6   Measures to Facilitate Changes in the Current System 86
Jenny Hunt and Heather Jackson

Part III Responding to the Challenge: Case-Studies of Good Practice — 95

Introduction		95
Chapter 7	An Adult Route to Qualifications *Louise Rowe*	98
Chapter 8	Managing Employment Training *Peter Boden and Sue Rose*	105
Chapter 9	Taking Education and Training to the Community: Providing Training Access Points in East Birmingham *Bob Addey*	111
Chapter 10	Supported Training in Employment for those with Learning Difficulties: The Manchester Experience *Joyce Thomas*	117
Chapter 11	Delivering Adult Compacts *Diane Harris*	124
Chapter 12	Guaranteed Accommodation and Training for Employment (GATE) *Claire Levy*	129
Chapter 13	Hitch-hiker's Guide to Science and Technology: An Access Initiative *Brenda Fulton*	134
Chapter 14	Employment Training Access Course *Ann Shoults*	138
Chapter 15	Community Joblink *Christopher McConnell*	142
Chapter 16	Tourism Innovation Project: An Example of International Collaboration *Lene Bak*	146
Chapter 17	APL/APEL and the Empowerment of Adults: The Centres Interinstitutionnels de Bilans de Competences in France *Armina Barkatoulah*	151
Chapter 18	Community Involvement and Development Course: Workers' Educational Association *Jol Miskin*	157

Chapter 19	Joint Initiatives of Industry and Trades Unions in Adult Education *Jacqui Bufton*	162
Chapter 20	The Rover Learning Business *Jenny Hunt*	171
Chapter 21	Meeting the Needs of Industry: The Westminster College Euro-Consultants Course *Gerry Smith*	175

| **Part IV** | **Summary and Recommendations** | **181** |
| Chapter 22 | Analysing the Past, Shaping the Future
Jenny Hunt and Heather Jackson | 183 |

| Bibliography | 187 |

Series Editors' Preface

Vocational education is now seen as an essential component of social and commercial policy and practice internationally. The recognition of the essential links between effective vocational education and economic and social development is resulting in a range of strategies and approaches within individual nations – and supranational bodies such as the European Community – to educate, train and retrain the current and future work-force to counteract increasing global competition.

In response, vocational education in the UK has experienced substantial change in recent years in terms of its curriculum delivery and organization. It has faced the implications of a demographic downturn and the need to extend the range of its provision, to meet the needs of both the adult unwaged and the new and expanding areas of professional, industrial and commercial updating. Britain is now responding positively to the challenge of the Single European Market and wider structural changes, both in Europe and across the globe, within a context that recognizes that, historically, the UK has failed by Western industrial standards to provide sufficient training to meet employers' needs. Measures are also being taken to significantly increase the participation rates for those aged 16 and over while also recognizing that some 80 per cent of the UK work-force of the year 2000 are already economically active.

The 1990s will witness the continuation and heightening of these processes and the generation of new opportunities and challenges for those responsible for providing vocational education. These opportunities and challenges will be mirrored internationally both in terms of the pattern of delivery and the balance of responsibility for action, varying between the individual, the employer and the state.

This series of books focuses on the opportunities and challenges facing vocational education. The aim of the series is to provide contemporary texts on emerging issues, with an emphasis on innovation and comparative analysis. More specifically, the series will seek to influence and inform practice by focusing on the application of new and original ideas. In this way, the texts will be at the interface between theory and practice with the explicit intention of enabling practitioners and managers to apply new educational ideas and philosophies. The books in the series are designed for those engaged professionally in vocational education and will provide information on, and critical analysis of, new developments. The contemporary and applied nature of this series will make it a valuable source of material on important current issues, building up into a library that illustrates the application of good practice.

Within this context, *Vocational Education and the Adult Unwaged*, edited by Jenny Hunt and Heather Jackson, makes an important contribution by focusing on the education and training needs of those adults without employment: it offers a powerful argument for developing a learning culture which enables individuals to realize their potential while supporting the development of a competent and capable work-force.

Through an extensive set of contemporary and relevant case-studies, combined with analysis, and recommendations of good practice, the book identifies a range of practical approaches to supporting the education and training needs of the adult unwaged which will be of interest and value to practitioners and managers. The book promotes a learner-centred approach to vocational education and presents a case for greater levels of resource support by employers and the state so as to encourage a culture of lifelong learning. As such, the book offers an agenda for future action in a sadly neglected area of vocational education provision.

We hope you enjoy this and other texts in the series.

Peter Funnell and Dave Müller
Ipswich 1992

Foreword

A book which places unwaged adults at the centre of strategic thinking about the education and training needs of the nation is both timely and welcome. So often, the needs of unemployed people have been addressed by short-term measures and narrow, job-specific training schemes unrelated to mainstream education and training policy.

There is now a widespread consensus among leading industrialists, academics and economists that current education and training systems cannot maintain the UK's status as an economic power. There is an urgent and overwhelming need to create a high-skill, high added-value, knowledge-based economy in Britain. Given that the work-force is undereducated, undertrained and underqualified compared to those in Europe and that, in the UK, 45 per cent of school leavers enter the work-force with no formal qualifications and that 80 per cent of today's work-force will still be in employment in the year 2000, there is a clear focus on the education and training needs of those in work as well as those who are unwaged. In general, there is a need to recognize the education and training needs of adults.

This book, with its focus on vocational education for the adult unwaged, not only addresses these issues, but also proposes strategies which will ultimately help to bring about the necessary changes for both individual and economic success.

Britain needs a multiskilled work-force with skills that are transferable between jobs and occupations. The development of a much stronger base of general competences, or core skills, is the way to flexibility and transferability. Britain must train to world-class standards. The existing divide between vocational training and academic education should be eliminated so that wider and deeper work skills that go beyond learning-by-doing are developed.

All of this implies a quantum leap in education and training. The message is lifelong learning. Changing attitudes to education and training among employers and individuals and the development of a 'learning culture' or 'learning society' is the challenge. The attitudes of people to training and development will be central to this process. At present, individual expectations are not high enough. People need support to plan their personal development. Priority needs to be given to developing people as well-rounded, flexible thinkers and communicators, and to valuing and taking into account their range of experiences of work and life.

This all requires learner-centred education and training, and new relationships between teachers and learners. Already, assessment and qualifications are being transformed and are rapidly becoming more accessible to a wider range of people. Education and training institutions are becoming more flexible and welcoming to adults.

What is needed now is an adequately resourced, national training strategy which embraces the needs of all individuals whether in work or out of it and wanting to join or return to the work-force. Employers, the state and individuals all have a part to play.

The case-studies in this book, along with the clear examples of good practice and the identification of key aspects of delivery which are important in the implementation of change, will all help those who are working with unwaged adults to move towards a new learning culture. They will also point us towards an achievable and much needed solution.

Lucia Jones
Chief Education Officer
Continuing Education
BBC

General Introduction

The development of the individual learner is now seen as an increasingly important aspect of adult vocational education and training, with less emphasis being placed on the idea of narrowly focused training for specific skills related to the immediate needs of the economy. This challenge of investing in people has the potential to create exciting opportunities for those out of work as well as those in employment, but it will take commitment, initiative and resources to make it work.

This book explores some of the issues currently facing people who are out of work, and will make recommendations as to how the new developments should take full account of the needs of the 'unwaged' and 'underemployed' as well as the employed, thereby enfranchising all adults into a set of national rights and entitlements to their own development and skills acquisition; a set of rights underpinned by a national commitment at both policy and practice level.

Jenny Hunt and Heather Jackson

List of Contributors

Addey, Bob Quality Improvement Leader, East Birmingham College of FE

Bak, Lene Educational Consultant, specializing in transnational programmes, working for Nelleman, Aarhus, Denmark

Barkatoulah, Armina Educational Consultant, specializing in APL and APEL, and working closely with the French 'Centres Interinstitutionnels de bilans de competences'

Boden, Peter Open Learning Co-ordinator, Chesterfield and North Derbyshire Chamber of Commerce and Industry

Bufton, Jacqui Education Officer (post-16 and adult), Gloucestershire LEA

Fulton, Brenda Course Co-ordinator, Sunderland TUC Unemployed Centre

Harris, Diane Adult Co-ordinator, Loughborough College of FE

Levy, Claire Educational Consultant/Trainer

McConnell, Chris Director, Finsbury Park Community Trust, North London

Miskin, Jol Tutor/Organizer, Yorkshire South District, Workers' Educational Association

Poole, Dennis Access Tutor, Brent College, north-west London

Rose, Sue North Derbyshire Chamber of Commerce

Rowe, Louise Adviser, Access to Assessment Service, North Lincolnshire College of FE

Shoults, Ann Access Tutor/Co-ordinator, Crawley College, Surrey

Smith, Gerry Adult Education Development Officer, London Borough of Wandsworth

Thomas, Joyce Senior Lecturer, Management Training Services Department, Tameside College of Technology

Woodward, Kath Organizer, Yorkshire South District, Workers' Educational Association

PART I

Introductory Perspectives

AN ANALYSIS OF THE CURRENT POSITION OF THE UNWAGED AND UNEMPLOYED WITHIN ADULT VOCATIONAL EDUCATION AND TRAINING, HIGHLIGHTING FACTORS WHICH DEMONSTRATE THE NEED FOR CHANGE

Individuals are now the only source of sustainable competitive advantage. Efforts must be focused on mobilizing their commitment and encouraging self-development and lifelong learning. (CBI, 1989)

Introduction

An official rhetoric is emerging from many different quarters – the CBI, the Employment Department, Training and Enterprise Councils – that acknowledges the need for a change in the UK's system of adult vocational education and training. It is at last being recognized that people are the key resource needed to generate the country's future economic wealth. As such, it must be in people that the major investment for the future is made. This official recognition of the importance of human resources has been a long time coming but, now it is here, everyone working in adult vocational education and training should be aware of what it means and how it could influence policy and practice.

The emphasis is on a 'learning culture', with a commitment from the Government and industry to developing and fostering a concept of lifelong learning in the adult work-force. In this context, generic skills will need to have the same value as specifically work-focused skills, and all adults have the right to ongoing education and training throughout their working lives.

If there is an acceptance that generic skills have a status equal to practical work-based skills then adults will, through increased opportunity for their own personal development, be better able to make informed choices about their continuing work roles. The development of a culture of continual and lifelong learning will bring greater flexibility and transferability both at individual and at industrial level. At present many adults who have skills which relate only to certain industries or economies are more likely to experience unemployment and, through a combination of limited skills and limited opportunities, are therefore not able to contribute their full worth to the economy.

There is a problem, however, with this new approach. While laudable in its content and rhetoric, there is a need to examine how it can be implemented. There is not much doubt that while the commitment to lifelong learning is to the advantage of all adults, it is still driven by economic need and not by a genuine desire to educate and empower the whole work-force. In this context, it is important to consider how this concept of lifelong learning can be fostered in a way that will truly benefit the individual adult worker and, more importantly, how the 'unwaged' adult will figure in these developments.

It is essential to provide unwaged people with the same opportunities and entitlements in adult vocational education as waged adults. Unwaged people should not be disenfranchised from a culture of lifelong learning simply because they are out of work.

Because it is the needs of unwaged people that are the focus of this book, Part I will start by:

- providing a definition of the term 'unwaged';
- providing an analysis of the current position of unwaged adults within adult vocational education and training;
- highlighting factors which demonstrate the need for change in vocational education and training for unwaged people; and
- beginning to explore further the argument for a holistic approach to vocational education and training for unwaged people.

1 The Unwaged and the Unemployed

A definition of terms

> Any attempt to respond to the educational needs of the unemployed has to recognize that, in terms of *range*, they are an extremely diverse group. Indeed, they share only one objective characteristic – the fact that they do not have a job. In other respects they are as diverse as the whole adult population in terms of their abilities, interests, needs, etc. In terms of balance, however, they include disproportionate numbers of people whose previous experiences in education have been unsatisfactory. (Watts and Knasel, 1985)

For the purposes of this study, our definition of the unwaged includes the registered unemployed as well as those groups who may not appear in the registered statistics but who are, nevertheless, not in work and looking for a job. While it is neither appropriate nor possible to identify a set of homogeneous needs for these groups they do, nevertheless, face some common barriers to 'employment'.

Attitudes to, and understanding of, the 'state of unemployment' vary enormously. Existing definitions of unwaged and unemployed cover a range of contexts, situations, perceptions and beliefs.

The UK Employment Department currently defines an unemployed person as 'any individual who is registered unemployed (signing on), and who is actively seeking work'. Anyone not formally registered in this way is not considered to be unemployed (by definition) and does not, therefore, feature in the unemployment statistics.

Even this definition of the 'registered unemployed' does not remain static. Over the past ten years the Government has constantly altered the way the unemployment statistics are collected, thereby dropping certain categories altogether from the figures. Changes have, for example, meant the loss of the following groups from the unemployment register and, hence, from the unemployment statistics:

- school leavers registered during June, July and August;
- married women without jobs but not entitled to benefits;
- men over 60; and
- adults on training schemes.

The benign interpretation of these changes in the way statistics are collected is that these people are not genuinely unemployed. The more cynical view is that different methods are constantly found for reducing the published unemployment figures. The constant changes in the way figures are collected could be construed as an overt political cover-up by a Government embarrassed by high unemployment.

Ironically, despite all these careful statistical reductions in the number of 'registered unemployed', the early 1990s have seen another real surge in unemployment. With the economy in recession during this period, even workers who were well qualified, and who traditionally had transferable skills in mobile sectors of the economy, have found themselves out of work. Nobody is immune to unemployment but a volatile economic scene is inclined to hit different groups harder at different times.

Groups hardest hit by unemployment

Contrary to popular perceptions, being 'out of work' is not some kind of incurable 'social illness'. Levels of employment are largely based on the health of the economy, and changing industrial infrastructures which redefine labour market needs. Lord Scarman in his report on the riots in some British cities in 1981 referred to the impact of technological change on labour market needs and opportunities:

> The structural causes of unemployment – which include remarkable developments in technology, an effect of which is that leading developed countries are losing the attributes of a labour-intensive economy – are deep and more complex than the mere existence of the current recession (1981). If this analysis is right, we shall have to face its implications. In order to secure social stability there will be a long-term need to provide useful, gainful employment and suitable educational and recreational opportunities for young people, especially in the inner-city. (Lord Scarman, in Jackson, 1985)

Lord Scarman's comments highlight the way labour markets respond to shifting industrial infrastructures. When this happens, there will inevitably be people who cannot respond to such changes, and who will therefore be more likely to experience unemployment.

Labour market changes, coupled with a range of other 'factors of disadvantage', put some groups far more at risk from short-term or long-term unemployment than others. The following unemployed/unwaged groups are frequently identified as those facing the toughest barriers to re-entry into the labour market:

- adults over 40, especially males;
- adults unemployed for over three years;
- unskilled and semi-skilled adults;
- adults living in rented accommodation;
- adults who have poor previous experiences of education and training;
- adults from black and ethnic minorities;
- women who have to care for more than one child;
- ex-offenders;
- adults with a history of mental illness;
- disabled adults; and
- adults with poor language and/or literacy skills.

Unemployment may affect people in different ways at different times. There is no doubt, however, that without constructive intervention – especially on behalf of the groups most at risk – a dangerous downward spiral of disadvantage, dejection and low self-esteem can ensue, making gainful, useful activity in the future increasingly unobtainable.

Clearly an individual's economic mobility and the range of options open to him/her will be influenced by a number of factors. The longer a person is out of work, for example, the more likely he/she is to stay out of work. The complex interaction of social, economic and individual factors that bring about the downward spiral damage both the individual and society as a whole. The waste in human resource terms for the economy is enormous.

For the sake of people experiencing the destabilizing factor of unemployment, and for the good of the wider society, thinking and practice must be developed to recognize the worth and potential of all adults and provide them with real education and training opportunities throughout adult life which enable them to play a full part in both economic and social affairs, both in and out of employment.

We have to move away from the idea that unemployed/unwaged people must be wedged into immediately available jobs – regardless of individual education/training needs and suitability – if we are to succeed in producing the high-quality, mobile work-force the economy needs and realize the full potential of the human resources of the UK.

Women and unemployment

> There are marked differences in the way men and women become and remain employed. These variations are the result of changes in male and female labour market segments, women's economic activity rates, employer recruitment preferences as well as women's propensity not to register as unemployed. (Charnley, McGivney and Sims, 1985)

Women have a special position in the labour market. Many of them do not bother to sign on and do not see themselves as unemployed in the traditional sense of the word. Nor do they qualify under existing regulations regarding unemployment benefit. Millions of women are, nevertheless, in the market for a job, as are the three million or so other adults who are involved in voluntary work or in a caring position in society which prevents them taking a traditional place in the labour market. All too often, these people find themselves in that marginal sector of the work-force – part-time employment. While in many ways this situation suits a large number of women because it fits in with family commitments, it can become a trap in itself.

Many women do not earn enough to provide for their families, but cannot find full-time work that is flexible enough to accommodate domestic commitments. Also, childcare is expensive and a vicious circle can emerge whereby women cannot return to work unless they get well-paid jobs which enable them to afford childcare. Many women become trapped in a cycle of dependency through part-time/low-paid employment and/or unemployment.

Ironically, during late 1989 and early in 1990, when the catch-phrase was 'skills shortages' (especially in the south-east of the country), women were being actively courted to return to the workforce. Flexible working patterns, home working and work-based creches were all being promoted because the economy required it. This happened despite consistent central Government rhetoric over the previous ten years that said women should be at home caring for their children. A new era appeared to be dawning, until the onset of the recession during the early 1990s. The economic arguments waned and, so it appears, did the initiatives.

The socio-political implications of being out of work

> Work may have different meanings for different people, but as well as these different meanings it also has an important shared meaning for people in society. Being a member of the work-force, enjoying the status of 'worker', leads to considerable positive reinforcement and satisfaction irrespective of the nature of the job. (Hayes and Nutman, 1981)

We have already stated that unemployed people are not a homogeneous group. Almost all unwaged people, however, experience some negative effects from being out of work. Research at the University of Sheffield (Warr, 1983) suggested nine such possible effects:

(1) reduced income;
(2) reduced variety of activity;
(3) few goals and less tractions;
(4) reduced scope for decision-making (choices);
(5) reduced scope for the development and practice of skills;
(6) increased exposure to psychologically threatening experiences;
(7) increased insecurity about the future;
(8) restricted interpersonal contact; and
(9) reduced social status.

It is possible to link these factors to the six main functions employment is deemed to perform in our society (Watts and Knasel, 1985):

(1) providing income;
(2) imposing a time structure on the working day;
(3) involving regularly shared experiences and contacts with people outside the nuclear family;
(4) linking the individual to goals and structures which transcend his or her own;
(5) defining aspects of personal status and identity; and
(6) enforcing activity.

There is no doubt that there are generally held social attitudes to 'having a job' which are clearly linked to social status, value and worth. Such prevailing attitudes make the imposition of unemployment a serious matter.

Probably the most damaging aspects of being without a job are the loss of self-confidence and self-esteem and the apparent closing down of an individual's ability to make choices about his/her life due to a lack of income. These lead to a loss of optimism about the future.

Michael Jackson in his book on Youth Unemployment (1985) refers to a statement made by the then Secretary of State for Employment, Jim Prior, at the time of the 1981 riots, on the danger that prolonged experience of unemployment could erode the 'ability to work':

> The long-term unemployed after a prolonged period of time are damaged by the psychological element attached to all this, and it makes it much harder for them ever to pick up work again.

The key to much of the debate around unemployment and its implications for individuals is whose responsibility is it? The implications of unemployment can be clearly identified, but solutions that do not recognize the underlying social, political and economic factors will inevitably fail to offer dynamic and radical solutions. This means that both the individual and society lose out.

The socio-political context

> Some people believe that unemployment is a short-term problem, which will eventually 'work itself out'. Others believe that it is a much more long-term problem, reflecting fundamental structural changes ... that are taking place in the nature and distribution of employment.
>
> ... In the end, however, unemployment remains a problem which requires economic and political solutions. (Watts and Knasel, 1985)

Any analysis of unemployment must take account of the macro/micro social and political context in which unemployment exists, the prevailing social attitudes to unemployed people, and the role and place of employment/paid work in society. It is important to make the definition specifically to *paid* work because there are armies of people engaged in voluntary/community activity and their work is not seen as 'real' since they receive no financial reward. Indeed, a central element of the attitude to employment/paid work is that of earning power and the status and privilege that stems from it. Women working as housewives and mothers have for generations been victims of this kind of social segregation – they do not 'earn', therefore they do not deserve the social standing and acceptance attributed to those earning a wage.

> In respect of the unemployed in general, Heinemann (1978) points out that worklessness is a socially described situation, and work is a social norm. Therefore, unemployment is seen as a fault of the individual. It is seen less as a

collective fate – a social, economic and political problem – and more as an individual failure. (Fiddy, 1983)

Writing in 1978, Heinemann identified an attitude to worklessness which reflects the attitude that if somebody is unemployed it is 'their own fault', and not the result of interlinking social, political and economic factors. This attitude to unemployment is not new. What has perhaps changed is the level of awareness and understanding of the economic reasons behind increasing unemployment. Some parts of the country have had high unemployment for some years now and, in these areas certainly, the attitude of 'he could get a job if he really wanted to' no longer prevails. Long-term unemployment in parts of the north, north-west, West Midlands and Wales especially has led to a necessary cultural shift away from the early stereotyped perceptions of unemployed people.

Even Government ministers acknowledge the structural nature of unemployment – for example, at the onset of the recession in the early 1990s, the then Chancellor of the Exchequer stated that 'a certain level of unemployment was a necessary evil towards economic recovery'.

Social awareness and understanding of unemployment may have lessened the stigma in some parts of the country, but this does not alter the impact of unemployment on the individual. A Government that accepts the structural reasons for unemployment must respond in more appropriate ways.

At present, Government policy and practice are unclear, inconsistent and largely unsympathetic, with frequent shifting of the way statistics are compiled in order to hide the political embarrassment of high unemployment, training programmes that still largely work on the ethos of getting unemployed people into the jobs available and continual changes in the regulations for benefit eligibility and benefit allowances. The Government continues to refuse to admit how serious the problem is in terms of the cost to the country and the potential for social unrest and discontent, particularly among young people.

In 1981 Lord Scarman reported on the riots that took place in many British cities. He drew attention to the link between high youth unemployment and other social deprivations in analysing the causes of the riots:

> The social conditions in Brixton – many of which are to be found in other inner-city areas – do not provide an excuse for disorder. They cannot justify attacks on the police or acts of arson or riot. All those who in the course of the disorders in Brixton and elsewhere engaged in violence against the police were

PART I: INTRODUCTORY PERSPECTIVES

guilty of grave criminal offences which society, if it is to survive, cannot condone ... At the same time, the disorders in Brixton cannot be fully understood unless they are seen in the context of the complex political, social and economic factors to which I have briefly referred. In analysing communal disturbances such as those in Brixton and elsewhere, to ignore the existence of these factors is to put the nation in peril. (Rt. Hon. Lord Scarman, 1981)

The reasons for unemployment are many and complex. Full employment never happens, but the level of unemployment is dependent on the economic buoyancy or otherwise of the country which, in turn, relies heavily on efficient handling of the economy by the government in power.

The political, social and economic factors affecting levels of unemployment are woven in a tight web of interrelated issues, with the unemployed person the fly in the middle.

We suggest, however, that, until governments recognize that unemployment is a waste of resources, with social as well as economic implications, and start to invest seriously in a culture of lifelong learning among adults, society will continue to waste its most valuable renewable resource, its people.

2 Adult Vocational Education and Training: What, Where and When?

> There is now little power in 'position', as your job can change in a minute; the sustaining power lies in what skills and ideas you have. (Kanter, 1990)

What is adult vocational education and training?

The answer to this question is that it is any education and training that enables an adult to identify and realize his/her potential, develop existing or new skills, and make informed choices about his/her role in the work-force and, indeed, in life in general.

It is not simply the acquisition of qualifications and skills for work. While its primary function is the promotion of the social and economic progress of society, it must also give individuals greater control, choice and power over their personal as well as their working lives.

> The education system is no longer merely a machine which reproduces the values and knowledge accumulated by past generations so that young people may be able to integrate rapidly and smoothly in a stable world which remains unchanged. Education has become a lifelong need, one of the best instruments for aiding and supporting individuals and social groups in an uncertain changing environment.
>
> Adult education for the long-term unemployed cannot provide the solution to the economic and structural problems besetting our societies. It can, however, make an essential contribution towards overcoming such problems: it guarantees a sound individual and collective perception of these problems and the preparation of suitable solutions.
>
> Adult education can facilitate participation in all aspects of life. It can help people to understand the society in which they live, give them the means to recognize and tackle the challenges of a society in which all aspects of life are changing, and rescue them from the processes of exclusion to which they have fallen victim. All sections of society need to understand the relevance of these comments.

PART I: INTRODUCTORY PERSPECTIVES

> Adult education is closely linked to the maintenance and development of social cohesion and democracy for which one of the prerequisites is reconciling the problems of employment and training. (Council of Europe, 1990)

Adult vocational education and training must be about a lifelong process of learning both in and out of employment. This learning covers a broad range of activities, not all of which are organized or intentional. A key aim for adult vocational education and training should be the support of the 'systematic acquisition of skills and attitudes for successful learning' (FEU, 1987), regardless of the class, gender, race or age of an individual. Learning opportunities must be available to all adults, whether employed or unwaged, on an equitable basis, and must recognize and respond to individual needs.

Traditionally, the view of vocational education and training is that of a period of preparation (school), followed by a period of training (either post-16 education or apprenticeship-type training at work) and then a period of action – ie, work. Inherent in this view is the belief that education prepares adults for their role in life and is a system primarily responsive to the needs of the economy. A manifestation of this thinking is evident in the school system up to 16, which is supposed to respond comprehensively to all individual, social, economic and cultural needs, to be topped up later by specific work-related skills training only as required.

This whole practice must be challenged and, indeed, has been over the years. The experience of adult educators has always been that adults grow, develop and learn in very different ways, at very different times, and according to a complex mixture of need, changes in life circumstances and personal development. There has to be a reflection of this in both national policy and practice, and a commitment to the entitlement of all adults to lifelong learning and education. Education and learning should be seen as a way of life, rather than an opportunity offered once with little chance of a second go if it goes wrong the first time.

If we approach adult vocational education and training in this way several fundamental changes could occur.

- It will be recognized that learning is a lifelong process and that the existing structures and delivery mechanisms must be modified to reflect this and ensure equality of access.
- The current panic to put vocational education and training earlier and earlier into school education would diminish because the prospect of a cut-off point for vocational education and training would no longer apply; this, in turn, could ease the pressure

on the school curriculum to prepare people once and for all for their working lives.
- Some of the inequities of the current system could be ironed out by providing clear access routes and opportunities for the development and acquisition of new skills throughout adult life and not just during school.
- Social mobility could be enhanced by providing greater opportunity for change in a lifelong learning framework.
- The notions of failure and success would be less finite. (Inevitably, a system that continues to have a very clear cut-off point marked by qualifications will succeed in rejecting large numbers of people.)
- There will be an increase in each person's potential for self-expression and realization of hidden talents.

> When lifelong education becomes a reality, it will be possible to offer greater scope to every person, to be less ruthless and tyrannical, and to provide for the needs of a greater diversity of people. (Lengrand, nd)

None of this is intended to negate the importance of early school years and the experience people have at that time. However, if a culture is developed that recognizes the rights of all people to equality of access throughout their lives, this may begin to reflect itself in current practice in primary and secondary education and, thereby, address some of the inequities at present built into that period of compulsory schooling.

Access to 'second chance' education

Access for adults into effective and quality adult vocational education and training is often offered as 'second chance', 'return to learn' or 'access' provision – the latter being an adult route into HE. Unfortunately, most of these opportunities have to be paid for by the individuals themselves, which makes them inaccessible to many adults who do not have adequate financial resources. Some excellent practice does exist in this area but it is patchy, frequently under-resourced, short-term and all too often carried out on the backs of enthusiastic and committed teachers and managers rather than through a coherent policy commitment. Much of the adult unwaged vocational education and training that falls outside the remit of Government programmes comes under this 'second chance'/'return to learn' heading.

PART I: INTRODUCTORY PERSPECTIVES

There is a philosophy attached to this whole area of work which focuses on the needs of people whose previous experiences of education have been poor, and for whom changes in life opportunities and choices seem to be increasingly unobtainable.

Central to much of the second chance approach is the belief that children from working-class backgrounds (and some black and ethnic minority children) benefit least from schooling and further higher education due to a cluster of factors of disadvantage, and are the least likely people to attend any form of adult education. They are also more likely to feel the effects of unemployment since they will probably be locked into unskilled and semi-skilled occupations which, in the current climate of economic and technological change, put them in a precarious sector of the labour market. They do not possess transferable skills and will therefore always be economically more vulnerable.

There is little doubt that school experiences alienate many people from systematic or institutional learning; this, in turn, limits their opportunities for personal and intellectual development. Above all, unwaged adults entering second chance programmes need to develop their innate ability, confidence and self-esteem. They also need to identify for themselves (with support and guidance) what their education and training needs may be and how they can best be met.

> The organization, curriculum and teaching methods used on second chance initiatives stem directly from a set of beliefs about how adults with minimal formal education can best be helped to develop their abilities, their critical thinking skills and their confidence, and about how to encourage personal autonomy both within learning and outside in the wider community. (REPLAN, 1988a)

The process of learning is of as much importance as the outcome, and both are enhanced when the learning is perceived as relevant by the learner and is based on a recognition of previous skills and experience. When people have been put off educational activity, it is very difficult to get them back. Once they find education again, they often discover a way of opening up their lives and developing skills which may have lain dormant for years, or which they may not have ever realized they had. In many cases, it may even be an opportunity to work on basic skills which were never acquired at school. In employment terms, all these approaches and experiences are absolutely vital if they are to re-enter the job market, or even find other ways of filling the time that they are out of paid employment.

There is no way in which education can 'solve' the unemployment problem. It can, however, attempt to respond to the needs of the unemployed. To some degree it can provide unemployed people with a structure and a sense of purpose. It can also help them to acquire new skills and knowledge. In these ways it can help them: to cope better with their existing situation; to escape unemployment and return to employment; and/or to use their situation as an opportunity to explore new possibilities. (Watts and Knasel, 1985)

The identified learning choices of an unwaged person may not necessarily be related to their immediate need for employment. They may also stem from other interests or activities, or even problems, they have, or they may just be areas and ideas that they want to explore.

It is important to note that, while the emphasis must be on the person and his/her needs in this whole area of work, any adult vocational education and training should not be approached solely on an individual basis. Programmes or initiatives aimed at the unwaged must take into account the social and economic isolation they often experiencce. The learning environment needs to provide unwaged people with knowledge of and contact with current labour market situations, as well as help to restore their confidence and self-esteem, and develop their skills.

There is considerable evidence from existing initiatives and the case studies in this book that suggests most unemployed people welcome a structure which enables them to get together with other people in the same situation as themselves. In this way different views and experiences can be aired and mutual support found, and time can be spent on identifying education and training needs and potential.

A variety of models

Many unwaged people feel that education and training have little relevance, especially if they are in a part of the country where there are very few jobs available.

Getting unwaged people into activities in which they can succeed and their ideas are accepted as valid is a first step to overcoming this resistance, and some outstanding examples of good practice exist. The Writers' Workshops in Liverpool, for example, were funded by the DES REPLAN programme in 1985 at a time when one in three of the adult males and one in five of the females in the city were unemployed, and where there was a growing recognition by those agencies working with these people that:

PART I: INTRODUCTORY PERSPECTIVES

> ... vocational training, although crucial, is only of limited value in areas such as Merseyside where there has been a collapse of the local economy and where the labour market cannot absorb those who have been trained – if it does so, it is often at the expense of other suitably trained people on the unemployment register. (REPLAN, 1988b)

The proposal was to use creative writing as a way of restoring self-confidence, putting unwaged people together in groups and, ultimately, changing people's negative attitudes about themselves and their abilities and potential.

The many and varied outcomes of the project related to the achievements and changes in the circumstances of both the people and the organization involved. The project developed, and was successful, in ways that surprised many of those concerned with its original development. It gained strength and momentum over its two-year life and its influence can be traced through a great many arts activities around Merseyside.

The emphasis in this whole area must always be that, being out of work does not make people redundant in themselves. The Liverpool experience is a shining example of what can be achieved when people are given time, support and encouragement in an activity to which they showed no previous particular ability or, probably, very little interest.

Work with unwaged adults must go further than narrow vocationalism. It must promote a broad curriculum in which a full range of educational options is advocated. This is particularly important in areas such as Liverpool and other parts of the country where vocational training offers very little opportunity of paid work at the end of it. The lessons learnt from projects like the Writers' Workshops in Liverpool demonstrate very clearly the value of this approach.

Watts and Knasel (1985) identify five key features of work with the unemployed:

(1) *Employability*: to help unemployed people to develop knowledge, skills and attitudes which will increase their chances of finding and keeping a job.
(2) *Coping*: to help unemployed people to develop knowledge, skills and attitudes which will help them to cope with being unemployed.
(3) *Context*: to help unemployed people to understand the extent to which the responsibility for being unemployed lies with society rather than with themselves and explore possible forms of social, political and community action related to unemployment.

(4) *Leisure*: to help unemployed people to develop knowledge, skills and attitudes which will help them to make good use of their increased 'leisure' time.
(5) *Opportunity creation*: to help unemployed people to develop knowledge, skills and attitudes which will enable them to create their own livelihood.

One of the most important issues surrounding work with unwaged adults is the recognition that whatever is on offer does not provide a solution to unemployment. Indeed, education and training plays a vital role in many respects but can only ever be seen as opening up options rather than offering solutions. Watts and Knasel consider the context of employment to be a key feature of work with the unemployed in that it offers the opportunity to explain and promote greater understanding of some of the realities of economic and labour market issues behind unemployment. By doing this, it helps people understand better the position they have found themselves in.

Ultimately, however, unemployment remains a problem which requires economic and political solutions. The lack of jobs with adequate pay and conditions is not an individual problem – it lies with the management of the economy as a whole. Good vocational education and training can only improve the potential of adults to move through the labour market more effectively, and to develop existing and new skills, or even to change direction. It cannot, however, change the structural reasons for large numbers of jobless people, nor society's attitude to them. Only political will and initiative can do that, and at present there seems to be a long way to go.

Government programmes – Employment Training

> We'll train the workers without jobs to do the jobs without workers. (Employment Department advertising)

In September 1988, the Government launched its new Employment Training (ET) programme for the unemployed. The scheme replaced all existing adult programmes and is still running, although it is now run regionally through the Training and Enterprise Councils (TECs) with a much reduced budget and an emphasis on responding to local and regional needs.

Employment Training was originally designed to *guarantee* skills training for all 18- to 25-year-olds unemployed for over six months, and also aimed to provide the same for 25- to 60-year-olds unemployed for over two years, enabling both groups to re-enter the job market. To achieve these objectives, the scheme combined elements of work experience with training activities, the intention being that, after one year's training, trainees would have a recognized qualification or credits towards one.

Unemployed people entered the scheme through a variety of routes, including a RESTART interview at the Employment Office. First referral was initially to a Training Agent for counselling, guidance, assessment and the preparation of training action plans. Candidates were then referred to a Training Manager who was responsible for delivering the programme of action as outlined in the training plan. It was intended that the training action plan should contain detailed objectives for each individual, covering basic skills or generic competences.

In some cases, where it was felt that the individuals needed some kind of training before moving into placement, a period of 12 weeks extended introduction was offered. This covered things like literacy and numeracy, computer literacy, communication skills, English for speakers of other languages, basic vocational training, additional counselling and job tasters.

Trainees received a 'benefit plus' allowance for being on the scheme, plus extra for fares, childcare and specialist clothing in some cases. Training agents received a payment for each trainee successfully placed with a Training Manager, and Training Managers were paid for each trainee registered plus an additional sum per week for each trainee on the books. Additional grants were also available in cases of high-cost training. Training Managers came to rely heavily on this last fund because the amounts paid per trainee were insufficient to cover the costs of delivering or subcontracting full-cost training.

Although the programme lacked the kind of commitment to adult entitlement, equality and lifelong learning we would like to have seen, and ignored the long-term nature of work with unemployed people, it did have some good points. For example, the emphasis on assessment, counselling, guidance and action planning was to be welcomed, as was the principle of sharing work-based activity with directed training.

It had two fundamental flaws, however, which prevented it from perhaps becoming a bench-mark in Government initiatives for the unemployed – it was underfunded and misleading. In most cases (and it must be said that there were notable exceptions, usually where additional subsidy was available to top-up inadequate Employment Department grants), Training Agents and Training Managers were unable to offer the time and space unemployed people required, or even respond in a genuine way to their needs, because of a lack of resources. Many could only exist by 'processing' a minimum number of people through the scheme each week, and the guidance and counselling needs of unemployed adults – or, indeed, any adults – cannot be satisfied effectively in this way. Unemployed adults frequently bring with them to counselling or guidance sessions an enormous set of urgent unmet needs, and these cannot be ignored if training is going to be effective and successful. Because of the poor resourcing, the scheme was not attractive to the professionals who have the skills to do the job properly.

In a survey carried out by the Centre for Alternative Industrial and Technological Systems (CAITS) on ET in London, 76 per cent of respondents recorded that it took less than a day to be assessed, and 43 per cent recorded that it took no more than an hour. In many cases trainees reported that they did not take part in a one-to-one interview, but were simply asked to fill out a questionnaire and sign the action plan!

Many Training Agents and Training Managers were small private training organizations with insufficient skills, expertise or time to really do the job properly. Good schemes were invariably those placed with organizations like NACRO, or attached to larger organizations and agencies which could subsume some of the costs.

As a result, it has proved very difficult to deliver ET in the way that was planned and certainly been almost impossible to meet the original aims.

Apart from resourcing difficulties, there also appeared to be a fundamental conflict in offering unemployed people training for work which was linked to their potential and skills when that work was probably not even available. The CAITS survey of ET in London recorded the following breakdown after the scheme had been running for nearly a year:

- 50,880 plans had been completed, resulting in just 18,988 people in training. 31,892 action plans (62 per cent) were surplus to requirements and six in every ten referrals failed to get a result.
- Around £1,335,200 was made from action plans alone.
- CAITS estimates that public funds of at least £557,000 were wasted on superfluous action plans in London during this period.
(CAITS, 1989)

PART I: INTRODUCTORY PERSPECTIVES

The scheme also fell notably short of targets in the first six months – 'out of the 300,000 expected to join ET nationally in the first six months, only 175,000 took part' (CAITS, 1989) – and, even where the outcome was successful, many trainees ended up in special projects, rather than work-based activities. The breakdown from the CAITS survey showed:

- 55 per cent of trainees were in projects;
- 32 per cent of trainees were based in a college;
- 4.5 per cent were based with employers;
- 4 per cent were in a project-college combination;
- 2.2 per cent were in a project-employer combination;
- 1.5 per cent were in a college-employer combination; and
- 0.7 per cent were in a combination of the above three.

This reaffirmed previous research carried out by CAITS which found that employers were not making a significant contribution to training or work experience. These figures also suggest that colleges had not successfully established effective links with employers. Project-based Training Managers were seen entering the market in competition with colleges, many offering accelerated training and claiming 'a standard equal to colleges and an ability to deal with a greater number of clients'.

Even more worrying, and the subject of much debate and concern, was the apparent coercive nature of the scheme. The CAITS survey found clear evidence of people being pressured into joining ET, either as a result of tacit compulsion or the threat of benefit suspension. Trainees and providers alike increasingly began to feel that the real agenda of ET was to get people into the jobs available as soon as possible so as to keep down embarrassingly high unemployment figures, and that there was very little real interest in individual education and training needs and potential.

Ironically, ET actually ran alongside the development of large areas of skills shortages in the country, particularly in the south-east, and the scheme made little or no impact on this situation. The intention was over-simplistic – 'to train the workers without jobs for the jobs without workers'. To have any chance of success, the scheme would have had to have been part of a much wider, longer term plan, with underpinning philosophies and commitments to adult vocational education and training, putting unemployed people on an equal footing with the employed.

There has been a substantial cut-back in the budget for ET for the 1991/92 period, reducing it by about one-third. According to the

Secretary of State for Employment, ET will be 'reduced in scale and reshaped' as a result of the new budget measures. The TECs, which are now responsible for delivering the programme, will have greater autonomy in the design and delivery process. It is interesting to note that some of them are returning to the traditional skills training model of previous schemes, like Training Opportunities (TOPs), and making their own decisions, within very tight budgetary constraints, about how the schemes should be delivered, and who should do it. A variety of developments is emerging out of this transfer of responsibility that appears to be an attempt to plug some of the bigger holes in the scheme. For example:

(1) Some TECs are looking at creating a consortium of foundation training providers better able to offer in-depth counselling for special needs groups than the Training Agent model.
(2) There is also some activity to encourage the Employment Service to undertake counselling, on the basis of fees paid by the TECs, for unemployed people who have some idea of the sort of training that they require. This would leave the consortium to offer more intensive counselling with a higher per capita investment by the TEC for unemployed people who have little if any idea of the type of training that they require.

If ET is really to emerge leaner and sharper then there needs to be a serious reappraisal of the philosophies and commitments underpinning it. Also, if the programme is not resourced in such a way as to fully respond to the needs of unemployed people, or if it is not based on a clear understanding and experience of their guidance, assessment and training requirements, then it will be little more than an elaborate job placement scheme with an unhealthy coercive element.

Officers responsible for the scheme at the TECs are only too aware of the difficulties facing them, and it will be interesting to reappraise the scheme in a year's time to see what has emerged.

3 Factors Influencing the Need to Change

Having considered existing practice in vocational education and training, other factors need to be examined which demonstrate the need for change. These are:

- skills shortages;
- rising unemployment;
- global competition;
- the demographic downturn;
- technological change;
- skills for the future; and
- 1992's Single European Market.

The skills shortage issue is one of the major factors which has forced a change of opinion about vocational education and training because it is so closely linked to issues affecting the unwaged and unemployed.

Skills shortages cannot be resolved merely by 'placing the workers without jobs in the jobs without workers'. In *Britain's Real Skills Shortages*, John Cassels (1990) says it would be easy to ignore the problem because statistics continually fluctuate from sector to sector and go up and down according to whether or not the country is in recession. The situation, therefore, never seems as bad as it actually is and, as a result, long-term responses are not developed. Also, as long as employers are able to 'poach' people who have been trained by other organizations, they will be tempted to place training at the bottom of the list of immediate priorities.

Cassels feels, however, that time is running out, and 'good luck' and 'getting by' are no longer good enough reasons to postpone really serious investment in and commitment to adult vocational education and training.

FACTORS INFLUENCING THE NEED TO CHANGE

The British economy in the 1990s has real problems. The CBI Quarterly Industrial Trends survey showed that, in October 1989, 19 per cent of firms were expecting skills shortages to limit output. A further survey conducted by the CBI on behalf of the Employment Department Training Agency showed that, in November 1988, nearly 61 per cent of firms were saying that their demand for skilled labour had increased over the previous year.

Global competition is now more intense than ever before, and the success of British firms in international markets will depend largely on the ability of their work-forces to perform as well as, or better than, those of their foreign counterparts. At the moment (see p. 40) they are not and the country is continuing to slip further behind European and international competitors.

Cassels also blames the poor quality of management and supervisory staff, and the low levels of investment and commitment to training, for Britain's failure to compete.

All indicators clearly point to the need for a better trained work-force at production, supervisory and managerial level, and one with transferable core skills which can be changed and built on over a period of time. All adults, whether employed or unwaged, should have the right to a minimum entitlement of education throughout their working lives. Commitment and investment at this level will make important inroads into the productivity of the UK as well as making a major contribution to the development of individuals and increasing the opportunities open to them. This is particularly important for the unwaged, for whom one of the main difficulties is a lack of opportunities.

The attitude that learning ends when people leave school has to be challenged, and a new culture which no longer talks about skills just in terms of training needs to be adopted.

The Chief Executive of Wearside TEC said at a recent conference about training that if skills were actually redefined in learning terms, then at least the root problem for education and training would be tackled. He described how the process should concentrate on making people enthusiastic and effective learners, encouraging them to identify their own goals and purpose. To date, many unemployed and unwaged adults have only been trained to do one job for a particular employer. Many skills and talents which could be put to productive use remain hidden, and many adults spend a large part of their lives in dull, monotonous work, with little opportunity for development or change, and even less opportunity to get another job should they become unemployed. Unskilled and semi-skilled workers feature disproportionately in the numbers of registered unemployed because they so often lack core transferable skills.

Other major factors which will bring about a shift in current practices are changes in production techniques caused by technological development. It is impossible to predict exactly what all these changes will be. What can be said with a fair amount of certainty, however, is that good generic, core skills which can continually be added to and developed throughout adult life will be essential in a work-force which will have to respond to rapid changes. Traditional employment patterns are going to be transformed and to ensure full opportunities are open to all adults – whether employed or unwaged – this will need to be recognized by placing a different emphasis on the role and place of the worker. It is being predicted, for example, that:

- the service sector will experience the highest growth, especially in leisure and tourism;
- transport, public health and education will also be growth areas;
- agriculture, mining utilities, manufacturing and construction will employ fewer people than at present;
- occupations with the highest growth rate will be management, the professions, personal and protective services;
- clerical/secretarial, craft and skilled manual occupations will show a slight decline; and
- semi-skilled and unskilled manufacturing jobs, including plant/machine operators and labourers, will suffer the greatest decline.

(Institute for Employment Research, 1989)

Also, in more general terms, there are some interesting facts about the future structure of employment:

- by the year 2000, early retirement will be much more common;
- only half of all paid workers by 2000 will be in full-time jobs, the rest will be self-employed, part-time or temporary workers;
- unemployment will still be with us (possibly three million);
- 70 per cent of all jobs in Europe will require cerebral rather than manual skills, which means that 35 per cent of 18-year-olds should now be entering higher education or its equivalent;
- 80 per cent of the present labour force will still be working in the year 2000;
- the bulk of new jobs will need people with higher level skills (managers, professional technicians and the like) not all of which can be created by traditional forms of higher education;

FACTORS INFLUENCING THE NEED TO CHANGE

- a common feature of all jobs will be the need for a solid grounding in basic competences such as personal initiative, enterprise, numeracy, technological, business and environmental awareness and communication skills, including fluency in other languages; and
- traditional job boundaries will break down, requiring people to become more flexible and able to diversify.
(IER Warwick, 1989)

Other more general predictions for the next century are more about things which will affect the levels and types of jobs and so, subsequently, affect how and what people will need to be trained for. For example:

- International competition is growing, particularly with the changing face of Eastern Europe, the opening up of the Single European Market, the rapid development of Pacific Rim countries and the dramatic recent developments in what was the Soviet Union. Britain, therefore, needs to at least keep up if not do better than its competitors by making the same kind of huge investment in learning as those countries which perform much better.
- Technological changes will affect every type of work-place and jobs at all levels. Information technologies will become increasingly integrated and, as new developments come on stream, managers will need to ensure that appropriate training is organized. Since traditional skill boundaries are breaking down there will be a widespread need to broaden training at all levels. Workers will often need new as well as traditional skills to use new technology effectively. Hence, there will be many job opportunities for those who can acquire new skills and job losses will be heaviest among the unskilled.
- It is predicted that there will be a sharp decline in the number of young people seeking work during the next decade, so strategies for recruiting and retraining mature workers and encouraging women to return to work, study and training will have to be developed. This will inevitably affect future methods of learning since these groups need a much more flexible approach to learning and one that will recognize their skills and experience.
- The advent of the Single European Market in 1992 will, according to the authors of *Vocational Education and the Challenge of Europe*, Peter Funnell and Dave Müller (1991), have an enormous influence on vocational education. It will add a European dimension to current ideas, plans and developments in this field.

PART I: INTRODUCTORY PERSPECTIVES

The European dimension

The Single European Market is probably one of the most significant changes in economic, social and political post-war European history.

It is generally agreed that the creation of a single market will bring with it new opportunities in economic and social terms. It will also have a considerable impact on the work-forces of member states, with an inevitable shift in the way economies are organized over the longer term and undoubtedly some restructuring and rationalization of production across Europe.

In this context, it is vital that members of the UK work-force – whether they be employed or unemployed – should have access to the same opportunities as those of other member states.

Read and Simpson (1991) suggest that, while the 'free market economy' and the 'contract culture' have become a part of everyday life in the UK, and have brought great prosperity to many people, they have also:

> ...required large numbers of people to be consigned to jobs characterized by low pay, no prospects, poor conditions, non-union and temporary, casual or unsocial in their hours of work. The large pool of unemployed labour during most of the 1980s allowed the whole question of skill and industrial training to degenerate to the level of 'Magic Roundabout' Government schemes, more about lowering wage expectations and 'parking' the unemployed out of the public gaze than about raising the skill base of the UK economy. (Read and Simpson, 1991)

The consequences of such an approach are now coming to light when the level of skills training in the UK is seen in comparison with that of other member states – the UK currently has 38 per cent of the population vocationally skilled, the lowest in Europe. Also, in Britain, six out of ten young people abandon education entirely at the age of 16; in France, Germany and the USA only one in ten do so. Comparisons between high-level skills training are also a cause for concern – Japan, Germany and France have twice the proportion of people undertaking higher levels of skills training and education as Britain, while the USA has three times the UK level.

One of the serious implications of this situation is that those people who have traditionally fared badly in the UK system of education, training and employment will do even worse in 1992, not better. The position of black people in the community is an area of particular concern. In many cases, black (migrant) workers from member states will not be entitled to the same free movement of other workers on the grounds of their citizenship status in the country in which they are resident. A combination of discrimination and disadvantage means that black people in the UK are already over-represented in the unemployment statistics; they are constantly dealing with discrimination when it comes to applying for jobs and are, statistically, the most vulnerable to redundancy. They are clearly not the only groups to be unfairly disadvantaged, but their position and status on racial grounds in a single market with confused messages about 'migrant status' is cause for great concern.

When he addressed the TUC Conference in September 1988, Jacques Delors, President of the European Commission, talked about the social imperative that must run alongside an integrated Europe, and raised five key issues:

(1) the freedom of movement of people across national boundaries (with consequent rights to live and work anywhere within the community);
(2) the establishment of a set of guaranteed rights – health, safety, terms and conditions – for people in their work-place;
(3) the provision of high quality training for the unemployed;
(4) the development of a regional strategy within Europe to tackle marginalization, underdevelopment and economic decline; and
(5) guaranteed minimum rights providing every citizen with an assured subsistence income, and defined rights for all in relation to health care and public services.
(Read and Simpson, 1991)

A more detailed version of these imperatives appeared in the 'Charter of Fundamental Social Rights' in 1989, which the UK Government has yet to agree to and which is the cause of much debate and controversy.

What is clearly happening, however, is that the European Commission is recognizing the need to develop an appropriate social infrastructure alongside economic integration to ensure individual rights and freedoms in a large economic market. In some areas there is concern that the charter does not go far enough because it is more concerned with the rights of 'workers' than 'individuals'.

This European approach to a supporting infrastructure for economic integration also manifests itself in recent education initiatives established by the Commission. The most recent, and probably the most interesting in terms of our debate, is the FORCE programme, which is based on point 15 of the 'Social Charter'. This states that:

> Every worker of the European Community must have access to vocational training and to benefit from it throughout his/her working life. In the conditions

PART I: INTRODUCTORY PERSPECTIVES

governing access to such training, there must be no discrimination on grounds of nationality.

The objectives of the FORCE programme are interesting and worth noting. They are:

- to encourage a greater and more effective investment effort in continuing vocational training and an improved return from it, in particular by developing partnerships designed to encourage greater awareness on the part of the public authorities, undertakings – in particular small and medium-sized undertakings – both sides of industry and individual workers, of the benefits accruing from investment in continuing vocational training;
- to encourage continuing vocational training measures by, for example, demonstrating and disseminating examples of good practice in continuing vocational training to those economic sectors or regions of the Community where access to, or investment in, such training is currently inadequate;
- to encourage innovations in the management of continuing vocational training, methodology and equipment;
- to take better account of the consequences of the completion of the internal market, in particular by supporting transnational and transfrontier continuing vocational training projects and the exchange of information and experience;
- to contribute to greater effectiveness of continuing vocational training mechanisms and their capacity to respond to changes in the European labour market by promoting measures at all levels and, in particular, monitoring and analysing the development of continuing vocational training to identify better ways of forecasting requirements in terms of qualifications and occupations.

It is also worth noting the goals of vocational training that underpin a programme like FORCE. These are:

- that of ensuring permanent adaptation to the changing nature and content of occupations and hence the improvement of skills and qualifications which is so imperative for strengthening the competitiveness of European firms and their staff;
- that of promoting social conditions, to enable large numbers of workers to overcome a lack of prospects for improving their qualifications and improve their situation;
- that of prevention, to forestall any negative consequences of completion of the internal market and to overcome the difficul-

ties arising in sectors or undertakings undergoing economic or technological restructuring; and
- that of integrating the unemployed, in particular the long-term unemployed.

The European Commission has identified the need to develop the appropriate education, training and social infrastructures to support a single European market, and to ensure all workers – whether employed or unemployed – have equal access to continuing vocational training.

Indeed, the messages coming from Europe are very much in line with the philosophy of this book, prioritizing as they do the need for policy, commitment and appropriate infrastructures to ensure equality of access to employment and ongoing skills training. There is talk of integrating the unemployed and ensuring that nobody is excluded from a set of fundamental rights for workers.

While the UK Government is willing to participate in EC programmes which provide funds for development, (eg, FORCE), it is still the only one not to have signed the European Social Charter. It is not imperative for member states to sign because the Commission regards such decisions as the responsibility of the Governments of individual member states. It seems an opportunity lost, however. While 1992 offers opportunities for some, it will close down opportunities for others. Restructuring and rationalization of economic infrastructures could bring about an increase in unemployment across the Community, with those groups always most vulnerable in unemployment terms worst hit.

The UK has to raise its skills base to compete effectively after 1992, and the Government is responsible for ensuring that all individuals have equal rights to opportunities for vocational education and training. The statements coming out of Brussels on continuing vocational training and the rights of both employed and unemployed people are a significant start. It is somewhat lamentable that the UK Government has declined to 'grasp the nettle' and move in this direction.

4 The Training Debate

... Britain's failure to educate and train its work-force to the same levels as its international competitors has been both a product and a cause of the nation's poor relative economic performance – a product because the education and training system evolved to meet the needs of the country as the world's first industrialized economy whose large, mass-production manufacturing sector required only a small number of skilled workers and university graduates; and a cause because the absence of a well-educated and trained work-force has made it difficult for industry to respond to new economic conditions.

The best way to visualize this argument is to see Britain as trapped in a low skills equilibrium, in which the majority of enterprises, staffed by poorly trained managers and workers, produce low quality goods and services. The 'equilibrium' is used to connote a self-reinforcing network of societal and state institutions which interact to stifle the demand for improvements in skill levels. (Gleeson, 1990)

The training debate has been escalating over the last few years, culminating in a deep concern about the quality, nature and structure of vocational education and training in the United Kingdom, both at youth and adult level.

The Government has responded to the problem with a variety of different initiatives and strategies which have already been referred to. The White Paper, *Employment for the 1990s* (HMSO, 1988), for example, contained a set of significant proposals which aimed at shifting the balance of control for adult vocational education and training away from the further education sector to employers. Six reforms were proposed to meet the changing economy of the 1990s, and increased global competition:

(1) training and vocational education need to be linked more closely to business and economic success;
(2) employers and individuals should accept a greater share of responsibility for training within a Government-devised framework;

(3) recognized standards of competence relevant to employment should be drawn up by industry-led organizations and validated nationally;
(4) training must provide access to qualifications based on these standards;
(5) responsibility for training should be devolved to a local level, bringing together private and public investment; and
(6) enterprises, individuals and local communities must be able to shape arrangements, programmes and opportunities to changing needs and circumstances.
(FEU, 1990a)

The main vehicles for these proposals have been the National Council for Vocational Qualification's (NCVQ) framework of competence-based vocational qualifications and the 82 regionally based Training and Enterprise Councils (TECs). TECs are employer-led organizations with a specific brief to provide local labour market intelligence, draw-up plans for securing quality training to meet local needs, and subcontract Youth Training (YT) and Employment Training (ET) and adult retraining initiatives.

Despite making clear statements about where responsibility for vocational education and training lies, the Government does not, however, define the nature of the provision, or apply any overall principles to the content and entitlement. In this respect it retains a largely 'hands-off' approach. The key to all recent developments has been the emphasis on vocational education and training being work-based, and, ultimately, employer-led.

There are clear advantages to be had from employers taking greater responsibility for education, but there are a number of areas for concern:

- While committed to guiding and advising employers to take responsibility for education and training and providing a framework in which this can take place, the Government is nevertheless reluctant to *oblige* employers to take action. This becomes a particularly serious issue with regard to small to medium-sized businesses, which actually make up the bulk of the employment sector and which are proving to be very reluctant to take responsibility for vocational education and training.
- The 'hands-off' approach adopted by central Government and the regional nature of TECs mean that there is no clear national delivery system, structure or curriculum for adult vocational education and training. For many adults the extent to which they

can benefit from ongoing education and training will depend on who they work for, or what part of the country they live in.
- There clearly needs to be a coherent national policy framework to underpin all adult vocational education and training in this country, and a commitment in both political and practical terms (resources must back up any such commitment) to the entitlement of all adults to ongoing lifelong education and training, and equality of access to such a system.
- Lastly, in an environment of employer-led vocational education and training, and the lack of any coherent policy or statement on adult entitlement, what happens to the unemployed? Where do they fit into the new training culture, a culture that places the emphasis for training on employers and responds to economic and industrial needs? The unemployed are clearly in no position to play an active part in such an environment.

Vocational education and training initiatives for the unemployed are currently delivered through a variety of outlets, funded by a range of different providers, and prioritized to a great extent by the organizations responsible for post-16 and adult education in local authorities. Government programmes also exist which aim to assist unemployed people to re-enter the job market. What all this provision lacks, however, is an underpinning belief that all adults have an entitlement to good, quality education and training throughout their working lives, whether they are employed or unemployed.

It is indisputable that the only real solution to unemployment is a job. It is vital, however, that adult vocational education training gives people core, transferrable and relevant skills, and isn't just a response to the immediate short-term needs of local and national economies. Good education and training undoubtedly empowers people in the labour market and, in turn, helps to promote a national culture which recognizes its importance and value.

Recent legislative developments

A statement like 'Training can no longer be a luxury for the minority; it is a necessity for all' is a clarion call, with a theoretical base reminiscent of the 1919 Ministry of Reconstruction Adult Education Committee Report which regarded adult education as a national necessity and something that should be both universal and lifelong. 'Thus the need for education and training for any individual will not be once and for all, but continuing'. (Small, 1984)

THE TRAINING DEBATE

A radical reform of the UK system of adult vocational education and training is required if the UK economy is to grow and develop, if Britain is to be competitive in Europe and wider global markets, and, above all, if the optimum use is to be made of the country's human resources.

The past few years have seen a number of developments which could have an enormous impact on adult vocational education and training. For example:

- the local management of FE colleges;
- the changes in LEA responsibilities;
- new funding arrangements for post-16 education and training;
- changes in training programmes for the unemployed with a substantial cut-back in funding;
- core skills developments, specifically for 16- to 19-year-olds
- National Vocational Qualifications (NVQs) and, more recently developed, General National Vocational Qualifications (GNVQs) (the latter have not been covered in this book due to a lack of precise information at the time of going to press);
- special funding for the unemployed who need basic skills; and
- the setting up of TECs with a brief to raise the standard and quality of training in their areas, and take on responsibility for schemes for the unemployed.

In some areas of the vocational education and training curriculum a coherent development can be traced – eg, standards-based education and training has its roots in experiments with a competence-based approach in early Youth Training Schemes, and developments have been fairly systematic ever since. The existing NVQ framework provides a formal structure for competence-based developments and accreditation in a work-based context.

In other ways, however, while some of the ideas have been intrinsically good, some appear to be based on political expediency rather than coherent strategic planning, and recognition of the importance and value of adult education in many developments has been sadly lacking. In fact, recent legislation about the changing role of LEAs and funding councils for post-16 provision has actually failed to recognize the needs of a very diverse adult market. Only intensive lobbying by the National Institute of Adult Continuing Education (NIACE) forced some rethinking and put the needs of adults on the agenda.

PART I: INTRODUCTORY PERSPECTIVES

The Further and Higher Education Bill

The reforms set out in the White Paper, *Education and Training for the 21st Century* (HMSO, 1991), which became the Further and Higher Education Bill on 4 November 1991 and became law in the 1991/92 parliamentary session, with major provisions taking effect from April 1993, had the following overall aims:

- to ensure that high-quality further education or training becomes the norm for all 16- to 17-year-olds who can benefit from it;
- to increase the all-round levels of attainment by young people; and
- to increase the proportion of young people acquiring higher levels of skill and expertise.

There has been considerable concern at the omission of an adult dimension from the original White Paper, and even greater concern about the impact of some of the legislation on existing adult provision. One of the reforms contained in the White Paper and the Bill – the independence of colleges of further education from LEA strategies, policies and accountability – will have a substantial and significant impact on existing adult vocational education and training provision because much of it is currently delivered through LEA provision.

Funding of local authority adult education and training provision has always been precarious, since the provision has never been a statutory obligation. This situation has been exacerbated in recent years in cases where local authorities, hamstrung by funding cuts, have cut the services that they are not in turn obliged to offer, adult education being one of these.

Under the Further and Higher Education Bill, the new Funding Councils will be expected to secure 'sufficient' full-time education for all 16- to 19-year-olds, but only be obliged to fund 'adequate' part-time provision for those over 19. Guidelines have been laid down as to the nature of 'adequate' provision:

- courses leading to recognized vocational and academic and vocational qualifications;
- courses approved by the Secretary of State leading to higher education;
- courses providing access to any of the above;
- courses providing basic literacy and/or mathematic skills;
- courses in English for speakers of other languages; and

- (in Wales) courses leading to proficiency or literacy in Welsh.

There is a commitment to some provision appropriate to adult learners, but there continues to be no systematic coherent policy or planning for the education and training of adults, and certainly no suggestion of prioritizing the needs of unemployed adults and people in other priority groups, who will often require special concessions on education and training before they can return to the labour market.

NIACE in its responses to the original White Paper made the following points:

> The aims of the paper 'Education and Training for the 21st Century' are (however) expressed exclusively in terms of developing the education and training system for 16- to 19-year-olds ... We believe that the consequences of the proposed legislation for adult learning have not been considered fully, and that whatever the White Paper's merits for the 16 to 19 age range, it does not yet represent an adequate strategy for adult education and training.
>
> Adult learners are not marginal. They are central to the country's economic regeneration. Most of the work-force of the year 2000 is already part of the labour market ...
>
> Almost half the adult population have no qualification of the old O level standard. NIACE welcomes the Government's commitment to improving this through increasing staying-on rates at 16 and believes that there is also an urgent need to better realize the educational potential of adults for the economy and the community. (NIACE, 1990)

The Bill, while recognizing the importance of adult learners in further education colleges, does not go far enough in acknowledging the range, scope and value of all forms of adult education and training, and the need for strategies to ensure a comprehensive, progressive system which can respond to a wide range of adult needs.

The whole adult dimension is sadly neglected in the recent legislation aimed at raising the quality of education and training in the 21st century and, while reassurances about the future of adult education are coming thick and fast from the mouths of ministers, there is still very little clarity and a continuing 'gap' in legislative priority and commitment.

National policy on entitlement and access

The ability to 'respond to change' has become an integral part of all work activities, and the need for a flexible work-force with transferable skills has become the clarion call of the 1990s. What must be borne in mind, however, is the fact that skills – transferable or

PART I: INTRODUCTORY PERSPECTIVES

otherwise – do not exist outside of people. It is the individuals that provide the skills and it is, therefore, individuals who must be put first and foremost when it comes to human resource planning and development. In an article in the *Times Educational Supplement* in 1987, Wellington emphasized the need for employers to recognize that:

- Skills do not exist in their own right. Employers do not recruit skills; they recruit people.
- Skills reside in people and are acquired by people.
- Skills are not entities which are in short supply. What industries need are people with the abilities to develop new skills, to learn new knowledge, to acquire new concepts and theories and to adapt to technological change with enthusiasm and lack of fear. (Wellington, 1987)

For industry to grow and prosper, there must be a continuing commitment to human resource development, because it is only this kind of investment and commitment that will ultimately result in a sound skills base.

The need to increase employers' commitment to vocational education and training is crucial, but it must be within the context of a national policy that ensures the rights and entitlement of all adults – whatever their economic or social position, and regardless of class, gender, race or ability – to vocational education and training, with clear lines of responsibility for delivery and recognizable, accessible entry points. Such a policy needs to underpin all initiatives related to adult vocational education and training, and ensure that all adults have access to education and training opportunities throughout their lives.

The omission of the adult dimension from the latest Further and Higher Education Bill, the apparent lack of concern about the disappearance of much of the existing adult provision delivered through local authorities, a refusal to comply with European Community recommendations on a social charter of rights and entitlements for adult workers in a European context, and a continuing lack of commitment to an underpinning strategy or policy framework to underpin the vocational education and training of adults mean that, in the near future at any rate, provision for adults will remain disparate, piecemeal, and dependent on who you work for or which part of the country you live in.

In this context, unemployed people will continually remain at the bottom of the heap, unable to take advantage of initiatives that are employer-based, and unable to afford full-cost provision, having only a limited choice available through 'increasingly obligatory' Government schemes for the unemployed. They will remain disenfranchised from real opportunities to develop their abilities and acquire new skills.

While the emphasis on encouraging employers to take an increased interest in training is to be welcomed, the inherent dangers of doing so without an underpinning policy or national framework of entitlement and rights to vocational education and training for all adults will mean the objectives can never actually be achieved.

The theory is right but, sadly, the implementation and process are short-sighted. Under the structures and initiatives currently being promoted, the majority of adults will never get an opportunity to take advantage of quality education and training, so it is unlikely that the skills base of the country will be significantly upgraded.

PART II

Vocational Education and Training Provision for Unwaged Adults

Introduction

One of the main characteristics of adult students that makes them different from young people is the diversity of their circumstances. Adults enter education and training with a vast range of personal situations and a wealth of experience. Their circumstances tend to be more complicated than those of young people because of their domestic responsibilities and financial commitments; also, they may be in full- or part-time employment, doing shift work or, if they are unemployed, they may be seeking work and need to be available for work at any time. In terms of educational background, the mature student's previous experience is often markedly different from that of younger students. They may have lost the habit of learning and, because of earlier school failures, doubt their ability to learn. All of these factors demand different kinds of provision, different teaching approaches and high levels of financial and personal support. An enormous amount of flexibility is required in terms of length, delivery and timing of provision as well as the ability to design programmes of work which are tailor-made to meet the identified requirements of the adult learner.

Access for unwaged adults

One of the best ways of encouraging adults into vocational education and training is to create an appropriate adult environment:

- rooms, buildings and tutors should all be easy to reach and find as well as being safe and accessible;
- timing should fit in, where possible, with the lives of adult learners and take into account things such as the school day, holidays and local transport;

PART II: PROVISION FOR UNWAGED ADULTS

- the learning environment should be welcoming and non-threatening;
- providers should do everything they can to ensure that unwaged adults know exactly what opportunities are available to them and ensure these opportunities are available to all adults regardless of their race, class, gender, age or disability;
- adequate and suitable childcare facilities should be provided;
- suitable refreshment and relaxation facilities should be available; and
- language and basic skills support should be available.

Finally, it is also important to remember and understand the enormous personal barriers to returning to education and training which many adults face. These actually prevent them from taking up opportunities. They are very real and can include:

- lack of confidence;
- no money for clothes, books and fees;
- pressure from friends and family;
- parental responsibilities;
- dependent relatives; and
- previous unsatisfactory experience of school.

Training and development issues for providers

The importance of meeting the needs of adults to encourage them to return to education and training has already been mentioned, but it is equally vital to invest in training the tutors who will be working with these adults if we are to implement change successfully. It needs to be recognized that the training and development of staff is a crucial factor in the creation and maintenance of morale and confidence among people whose work with adults demands considerable personal investment and commitment. Staff who have been used to traditional, formal teaching and training methods, or working specifically with younger people, will need training opportunities to develop new skills, adopt greater flexibility and respond to new demands. They need to understand the place of the following in work with unwaged adults:

- education and vocational guidance and counselling;
- the negotiation of learning objectives;
- the range of learning styles and materials available for adults;
- accreditation possibilities;

- the development of the process as well as the content of learning;
- monitoring progress and performance;
- personal and organizational barriers to adult learning;
- competence-based learning;
- collaborating and negotiating with other education and training providers and employers;
- initial assessment; and
- flexible delivery methods.

It is not just those staff involved in teaching and training who should benefit from updating and development. Everyone dealing with adult students, from administrators, reception and support staff to caretakers and cleaners, must learn how to adjust their roles and responsibilities in response to new developments and so understand, contribute to and benefit from how education and training is delivered in their particular organization.

It is extremely important to give staff proper training instead of expecting people to 'pick up' new skills and roles which are required of them. To ignore the training needs of adult providers is to marginalize adult learning. Given the stated needs of the economy and the adult learners themselves, training and development must be a priority if the best service possible is to be offered and a better learning environment for adults is to be secured.

Probably most important of all, providers of adult vocational education and training need to change from being teachers, in the traditional sense of the word, to facilitators of learning.

Quality and equal opportunities

Quality of provision and equal opportunities are enormous subject areas, each justifying a book in their own right. Their importance in terms of the creation of appropriate provision for all adults, however, cannot be emphasized enough. All adults must, regardless of class, gender, race, age or disability have equal access to the opportunities offered by vocational education and training. Courses and training schemes must be designed so that they are open to *anyone* who wishes to participate.

Ensuring that the adult students who participate in educational and vocational training are taught in the best possible premises, by the most experienced staff, who are delivering appropriately designed courses using the very best materials and equipment, is key to the kind of changes which are being advocated.

To ensure that this happens, effective methods of measuring delivery, performance and resources must be established, remembe-

PART II: PROVISION FOR UNWAGED ADULTS

ring of course that qualitative evidence is as important a measure of success as quantitative evidence. The evaluation of a service for adults needs to incorporate information about provision in terms of its fitness for purpose, ie, methods of delivery, content of programmes, appropriateness and relevance, interests and expectations of client groups and individual learners.

5 Six Key Aspects in the Delivery of Adult Vocational Education and Training

In terms of opening up and improving vocational education and training opportunities for unwaged adults, there are six key elements which must be incorporated:

(1) educational and vocational advice and guidance;
(2) teaching and learning methods to suit adult needs;
(3) an improved system of financial support for adult vocational education and training;
(4) the accreditation of prior learning;
(5) developing competence-based learning through NVQs; and
(6) developing and embedding a core curriculum.

Each of these approaches will be dealt with by:

- providing a definition and explaining the mechanisms, where relevant;
- describing the benefits to be gained;
- identifying development implications; and
- pointing out some of the problems.

1. Educational and vocational advice and guidance[*]

Educational and vocational advice and guidance enable people to make realistic choices about education, training and employment possibilities. In order to make valid and informed decisions in these

---

[*] The authors wish to thank Dennis Pools of North West London College for all his work on this section of the text

areas, people need access to information and advice which can help them identify their existing skills and potential and recognize where there is a need for further training, and which give guidance on available opportunities.

This is an essential process for unwaged adults because it gives them the opportunity to reappraise their existing skills, to think about a wide range of possibilities in terms of training and work, and helps them to consider ways of overcoming unemployment and making significant changes in their lives.

At a time of shifting work patterns, changing skills needs and increasing globalization of markets, it is vital that unemployed/unwaged adults can engage in this type of process. Not only does it address issues of individual self-esteem and self-confidence; good education and vocational advice and guidance can add 'new' resources in economic terms to the country's wealth by identifying potential for retraining and multiskilling.

In this way, effective guidance also meets the needs of employers. Employees, or prospective employees, who have received advice and guidance and consequently chosen a job because it genuinely suits them, who have had appropriate training and who are aware of the possibilities for further development, are likely to be better motivated and more effective than employees who have not been involved in an advice and guidance process.

Mistakes in employee placement are just as costly to the economy as large numbers of unemployed/unwaged people. The guidance process can perform a brokerage role, ensuring the best matching of the needs of the individual as well as the needs of the employer.

Advice and guidance also helps people to find their way through what is frequently a confusing and intimidating range of education and training provision, with a variety of different access points and entrance criteria. This confusion and intimidation can only be resolved by providing some form of coherent educational and vocational guidance.

Why is guidance so important?

Guidance is obviously important to people who are seeking practical outcomes to enable them to function more effectively as human beings and as workers. Unless guidance can offer relevant and realistic outcomes, it is failing. The effectiveness of these outcomes must be matched by sufficient progression both in terms of appropriate provision on offer and appropriate and available employment. In this sense, guidance becomes a key interface between the needs of people and the economy, and the means of obtaining appropriate skills and qualifications to match them.

Guidance not only offers solutions and helps individuals realize their potential, but also acts as a catalyst for change. Feedback to education and training providers on the needs identified in the advice and guidance is a key resource in the development of relevant and appropriate provision.

Guidance is not therapy, it is not palliative and it is not peripheral. It is a vital means of ensuring that whatever choices people make have realistic and practical outcomes. It ensures that the economy and, ultimately, society benefit from a well-trained and motivated work-force.

What does effective guidance involve?

Effective guidance has to take full account of the individual needs of the person seeking it. It is measured primarily by the efficacy of the outcomes arising from it. The Unit for the Development of Adult and Continuing Education (UDACE, 1986) identified seven activities that guidance needs to address: (1) informing, (2) advising, (3) counselling, (4) assessing, (5) enabling, (6) advocating and (7) feedback. These can be achieved through group or individual participation in the guidance process, or by guidance agencies networking together to share knowledge, experience and expertise.

The appropriateness of each activity mentioned above depends on local and individual factors and the (mis)match between them. It is here that it is possible to identify readily the role of guidance as an interface between individual needs, the needs of the economy and the types and relevance of education and provision on offer.

Individual longer term learning outcomes of guidance are more difficult to define. Killeen, White and Watts in *The Economic Value of Careers Guidance* (1991) recognize the learning outcomes shown in Figure 1 as 'closely associated with the process of guidance'.

In their work, Killeen, White and Watts concluded that:

> ...positive results are documented for all major guidance strategies across most learning outcome types, and that the effects of successive guidance interventions on learning outcomes can be cumulative.

Guidance and unwaged adults

Many unwaged adults who seek vocational or educational guidance do so because they are unsure of options or because they feel intimidated by what they see as elitist educational institutions. It must be recognized that those seeking guidance are often in a

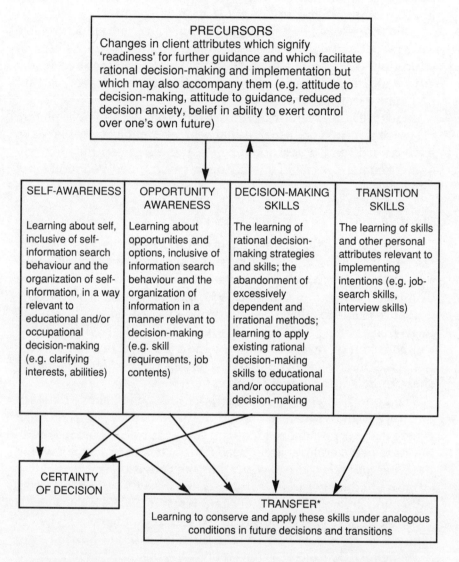

* The final outcome category conforms with current conceptions of good practice and with the general philosophy of learning outcomes. Evaluation studies of learning transfer have not yet been reported.

Source: Killeen and Kidd, 1991
Classification derived from Law and Watts, 1977

Figure 1 *Learning outcomes and closely associated outcomes of guidance*

vulnerable situation and lacking in self-confidence. Being out of 'paid work' itself can eat away at self-esteem and self-confidence; in particular, it can bring financial problems and many people, especially women, may have expensive childcare responsibilities. Some people may feel they are victims of racial, gender or age discrimination. Still more may have had negative experiences of previous educational provision.

All in all, many unwaged adults seeking guidance feel marginalized and disempowered. Effective guidance needs to, or at least start to, overcome such feelings. All guidance provision with unwaged adults must be easily accessible, offer confidentiality to enquirers, be non-threatening and genuinely seek to build confidence through recognition of existing skills and competences.

How is guidance delivered?

Information, advice and guidance are delivered by a wide range of agencies, including:

- Careers services (where they are resourced to work with adults);
- Educational guidance agencies (where these exist);
- Local education offices of LEAs;
- Educational institutions;
- Training providers;
- Employment service;
- Community-based organizations;
- Voluntary sector organizations;
- Libraries;
- Employer-based activities;
- Private providers; and
- Home office provision.

Careers offices, where they are resourced to work with adults, offer easy access to information, advice and guidance since they are usually found on the high street. Education guidance agencies also often have a shop-front location, although funding cuts over recent years have caused a serious decline in this type of provision. Colleges are increasingly developing central admissions services, to which all enquiries are referred for information, advice, and, if necessary, guidance.

Libraries are regularly used by local people as important information and advice providers, but not for guidance. Education enquiry offices in LEAs are frequently on the unsolicited end of numerous and wide-ranging guidance requests. And a large number of voluntary sector and community-based organizations are the only source of advice and guidance available to many adults.

The importance of guidance networking cannot be emphasized too much here. Clearly, given such a wide range of agencies, each one having an important part to play in the guidance process, it is crucial for each of them to know about the others, understand their place and position in the 'guidance process' and be fully aware of who to refer individual enquiries on to.

Working with unwaged adults generally imposes, on all of those engaged in it, the need to network with a wide range of agencies. Adults, particularly those who have experienced employment, bring with them a mass of information and needs, some of which are clearly educational and vocational and others which are not. The ability to distinguish between them, identify the priority needs, and refer individuals on to appropriate agencies is a key guidance skill, and one that emerges from effective networking between all the agencies.

Guidance networking

Guidance networking is about agencies involved in the delivery of guidance working together to provide a better, more coherent and more accessible service to individual adults. It requires a commitment by the agencies concerned to sharing, collaborating and cross-referencing, and is the only way a truly effective, systematic guidance service can be developed. No one agency can contain all the information and expertise necessary to respond to the needs of all adults.

TECs in many parts of the country are playing a strategic role in the setting up of guidance networks. The demise of LEA control of post-16 education has meant that the TECs are well placed to take on the facilitating/brokerage role in a guidance network. They have an overview of provision in their areas, are committed to effective guidance underpinning their own training programmes and, not being in a delivery role as such, are able to retain an objectivity in a guidance networking function.

Can the effectiveness of guidance be measured?

In an educational ethos that is increasingly emphasizing performance indicators and quality control, it is important for guidance services to be able to monitor and evaluate their provision. Much of the quality control in guidance will revolve around the skills and competences of staff and the systems in place to deal with enquiries.

Applying performance indicators to individual outcomes from the guidance process is almost impossible given the enormous range of individual needs and circumstances that adults bring to the process and the buoyancy or otherwise of the job market.

There is no doubt, however, that the development of effective systems for the delivery of guidance that produces positive outcomes for the enquirer is of crucial importance, as are the competences and skills of staff involved and the quality of the networking activities guidance agencies are willing to engage in.

How can guidance be adequately funded?

Most guidance services to date have been funded through local authorities or through central Government initiatives and agencies. The major agencies – such as the careers services and educational guidance agencies – have traditionally been run by LEAs, and much of the guidance that is undertaken in adult retraining schemes has come via the Employment Department programmes.

There are now obvious threats to the funding of educational and vocational advice and guidance through this traditional route. Guidance activities have always been non-statutory and, as such, are at risk during times of cuts and changes.

It is clear that TECs will play an increasingly significant role in the support and promotion of adult education and vocational guidance, particularly with the administration of the Government's new Gateway programme, which emphasizes the need for effective, coherent guidance services across TEC areas, and promotes the use of 'guidance vouchers' for adults.

Educational and vocational guidance is too important to be left to the market-place

The argument for coherent, well-resourced guidance provision has been on the education and training agenda for many years, but it continues to be on the fringes of provision. Experience has shown that there is an almost unlimited demand for educational and vocational guidance. During the 1992 BBC Second Chance initiative, which was linked with Adult Learners' Week, for example, the Employment Department special help-line dealt with more than 57,000 enquiries. As economic circumstances change, this sort of provision is going to be even more in demand.

People on the dole queue lose their self-esteem and become alienated from society and its goals. Unemployment continues to be a serious drain on the nation's wealth, and is a complete waste of human resources as far as the economy is concerned.

Educational and vocational advice and guidance are key services in helping unemployed/unwaged people onto education and training programmes, back into the labour market or, even, to realize and develop new skills. Any strategies that facilitate changes in people's circumstances and build self-esteem and self-confidence must benefit the economy generally.

To restrict funds to guidance agencies, and allow them to be lost in the mass of cut-backs being experienced as a result of changes in LEA responsibilities, is short-sighted and self-defeating.

Commitment to, and funding of, educational advice and guidance for unwaged adults is central to the development of the people concerned as well as to the long-term health of the economy.

2. Teaching and learning methods to suit adult needs

If there is to be a 'revolution' that will radically alter the basis of vocational education and training, then there also needs to be a change in the way education and training is delivered. Only then will there be a system which is appropriate for, and responsive to, its adult clients and which takes into account the characteristics of adult learning. What will be described now are ways in which teaching and learning styles and methods of delivery can be changed to enable providers to achieve that aim. What is being referred to specifically now are the ways in which a course can be *designed*, *delivered* and *organized* so that a user model is established rather than one that is provider led.

Scheme design and development

Very often, adult learners, and in particular the unwaged, need to have some formal recognition or credit for what they have been doing on a course after a relatively short period of time. They may get a job or their personal circumstances may change and so affect their ability to continue. To prevent any sense of failure or feelings of having wasted time, it is important that all educational and training courses are designed on a modular or unit basis. Such a method allows students to stop and start their studies without being disadvantaged in any way because the course has been broken up into small separate blocks or accessible units which do not necessarily depend on sequence. This allows students to pick and choose, mix and match, and stop and start, according to their particular needs at any one time.

It is important here to note that the successful delivery of a modular course requires an increase in the amount of guidance and counselling available to the adult learner. Educational guidance staff would be needed to assist students in the selection or negotiation of appropriate modules. The benefits of this form of modularization for adult learners include:

- increased choice and access to the full range of provision;
- improved opportunities for credit transfer and progression;
- increased viability in terms of numbers on courses in some areas of provision;
- increased flexible learning opportunities;
- increased opportunities for cumulative credit;
- recognition of prior learning and achievement; and
- increased access to learning opportunities.

Also, in terms of the design of courses, it is important that colleges in particular abandon their traditional practice of a September start. Unwaged adults need to be able to enter programmes of learning and training at different times and roll-on and roll-off courses and schemes according to their individual circumstances.

Most importantly of all, for the unwaged learner, modularization has the distinct advantage of not separating their learning from the mainstream. It provides institutions with the potential for moving educational and training opportunities for the unwaged from the margins to the mainstream because it allows young and old, the employed and the unemployed, all to work together.

Delivery of teaching and learning

The introduction and implementation of a modular system inevitably requires a flexible delivery approach which is substantially different from the more traditional methods. What is now needed is a wide variety of delivery styles, ranging from traditional teacher-directed instruction of groups to totally open provision which allows learners the freedom to design their own learning programme and pursue it at their own convenience. In other words, we need a more individualized approach which is extremely beneficial to the adult learner.

These new methods of delivery include:

- distance learning;
- computer-assisted learning;

- work-based or work-place learning;
- open access workshops;
- independent study with peer-group or tutor support; and
- open learning.

The organization of teaching and learning

This is all about how tutors actually do their work with their clients. It requires an experiential, individualistic approach rather than one which is too theoretical and dependent on note taking. It incorporates discussions/debates, workshops, simulations, case-studies, research, lectures, projects, group work, discovery methods, practical work, sharing experiences, and demonstrations.

To be successful, this way of working requires: good, caring student/staff relationships; a friendly relaxed atmosphere; furniture arrangements which allow effective interaction; furnishings which are attractive and adult sized; a team spirit between students and teachers as joint enquirers; and an atmosphere in which people can express themselves freely without fear of ridicule. These are all conducive to learning.

The benefits of working with adults in this way are numerous because: learning becomes interesting and relevant; clients are well motivated and therefore there are low drop-out rates; everyone has the opportunity to contribute and participate; clients have the opportunity to influence the direction and content of courses; and clients can use their own experiences as resources for learning.

The development implications in both financial and staff development terms for what has just been described are enormous and a huge commitment is being asked for when it is suggested that the following ideas are implemented:

- organizations will have to completely change the way they operate;
- staff attitudes will have to be altered through extensive staff development programmes; and
- methods of instruction will have to be adapted to encourage learning autonomy and interaction.

It is encouraging to see how much of this has happened already, but there is still a lot to be done. How else can unwaged adults be encouraged to enter and stay in a system of learning and training? Efforts must be made to change approaches so they meet the needs of students rather than just those of the organization and those who work there.

SIX KEY ASPECTS IN DELIVERY

3. An improved system of financial support

A section about finance may seem to be inappropriate in a book concerned with a philosophy which focuses the attention of providers of education and training for unwaged adults on the development of the individual learner and a lifelong learning culture. Although there is not necessarily a direct relationship between the amount of money spent and the actual quality of provision and delivery because very often a great deal of what has been described can be achieved without enormous financial inputs, something which is crucial to that philosophy's total implementation cannot be ignored. This section is here to remind practitioners about the kind of financial support which they, their students and their organizations need to create a proper learning environment. Such information can empower tutors when they are requesting further resources.

It is not easy to ask for and obtain additional funding. To clarify what to ask for and generally assist the process, the following aims should be borne in mind. What is needed is increased learner support, increased staff support, increased institutional support and increased Government support.

Increased learner support

Ideally, those unwaged adults who are returning to education and training should not have to struggle unnecessarily in personal terms with their finances, because learning new skills and coping with a change of circumstances is difficult enough without this particular worry. More and more evidence is being produced to reinforce the argument that personal financial problems are a major obstacle in terms of returning to learning. Therefore it must be a priority to find ways of providing unwaged adults with enough money. Pressure needs to be exerted so that:

- grants, loans or pay-as-you learn schemes for adult students become an entitlement so that course fees and general living costs can be met;
- unemployment benefit is not affected by the amount of time spent in learning – people should be rewarded, not penalized, for their efforts;
- schemes such as free bus passes, transport vouchers, childcare allowances and meal vouchers are introduced;
- education and training passports are developed which allow for full fee remission on courses;
- training credits for adults are introduced; and

- community care schemes are developed for the care of elderly or disabled relatives while adult carers are learning.

Increased staff support

What people responsible for the education and training of unwaged adults need to know and change has already been identified. What is also needed is a commitment from employers to the development of their staff. Creating a well-trained and valued work-force, with appropriate levels of resources and support to work with unwaged adults, is a key factor in successfully implementing the changes being advocated. So, financial support is needed for:

- staff training to update skills and learning about new teaching methods and types of assessment;
- equipment and resources which enable providers to keep up to date with developments; and
- staff cover so that people can be released for networking activities, employer liaison, curriculum planning, project work and outreach.

Increased institutional support (with specific reference to colleges)

Statistics demonstrate that more and more adults are entering colleges for further education and training. This demographic shift means that these institutions need to give a higher priority to provision for this particular client group and make creative changes to policy and practice which will enable more appropriate courses to be established and run in a more flexible way for adult learners. For example, colleges must recognize that attracting adult learners can be more expensive than providing initial vocational training for young people because money is needed for resource-based learning, guidance and counselling, distance learning, the adaptation of buildings, outreach (staff and premises), new assessment systems (APL), and new qualifications in addition to the usual tutor costs. However, such developments could be seen as being beneficial to *everyone*, not just adults. The long-term benefit would be that people would want to return to a system of learning which was more accessible, more attractive and which catered for their individual learning needs. Colleges which are committed to the development of education and training for unwaged adults should also be considering the financial benefits of introducing adult compacts, income-generating activities, customized training and self-financing courses so that surpluses in one area of college work can subsidise others.

Increased Government support

All of these suggested changes depend a lot on the individual organization's perception of its role as a provider and there is no guarantee that they will all respond positively to what is being proposed. To prevent an even more piecemeal pattern of provision developing than already exists, particularly in the light of the most recent Government White Paper which will make all FE colleges independent bodies by 1993, there is an obvious need for national guidelines and legislation to guarantee funding which will allow greater participation in learning and training by unwaged adults. For example:

- institutions could be rewarded for recruiting these non-traditional students;
- steps could be taken to convince policy-makers about the importance to the economy of encouraging greater participation rates by adults;
- there could be stronger incentives to stimulate employer investment in and support for, employment and training; and
- collaborative ventures could be encouraged in order to maximize the resources available.

But, above all, in terms of increased Government support, what is needed is an overall political commitment to the whole idea of developing a lifelong learning entitlement for adults. Government investment in vocational education and training needs to rise dramatically to match the increase in individual motivation and institutional commitment which is already taking place. There is no other easy answer. The Government must realize that, if the needs of our economy are to be met by a better trained adult work-force, more cannot be achieved with the same or less.

4. Accreditation of prior learning

A definition of terms

APL: Assessment/Accreditation of Prior Learning
APEL: Assessment/Accreditation of Prior Experiential Learning
APA: Assessment/Accreditation of Prior Achievement

A variety of acronyms is used to describe basically the same thing, namely the process of identifying, assessing and accrediting competences which a person already has.

What and Why

It is now generally accepted that people learn and achieve levels of competence throughout their lives, and not just through the traditional learning routes of schools and colleges, or formal qualifications.

The process of assessment and accreditation of prior learning (APL) seeks to identify and accredit the knowledge, experience and skills people gain throughout life. APL enables people to recall, reflect on and define learning and achievements derived from life experiences whether they be at home, work, leisure or through community or voluntary activities.

A range of outcomes can emerge from this type of analysis and identification, some direct and others indirect:

Direct outcomes
- the identification of previously unaccredited work skills and experience;
- the identification of skills and experience gained outside of work;
- the identification of competences for the purposes of NVQ accreditation; and
- the identification of a basis for an individual development plan.

Indirect outcomes
- reflection on past and current activities, and the review of goals;
- building confidence; and
- opening up opportunities not previously considered.

As a process that helps to identify and value people's skills and expertise, even where this has not been previously formally recognized, APL is invaluable. It is particularly suitable for those adults who have more experience and a higher level of skill than their paper qualifications would suggest. In this sense, as a tool of individual 'empowerment' it can be seen as especially significant.

In France, about 80 centres of 'bilans de competences' have been set up with a view to enabling adults, through an APL process, to review their current position in both economic and personal terms, thereby empowering them to make plans and effect changes in their lives.

In the UK, too, funding is being made available to the 82 TECs for setting up 'expert teams' with a view to raising the number of people who have access to NVQs through the APL process.

The process

Providers of education and training and employers will need to develop very specific skills in order to deliver APL effectively. Even recruitment to the APL process can make challenging demands on existing practices and resources, and has to be addressed in non-traditional ways. Very few adults not currently involved in education and training understand APL.

Assessing competence in unpaid work

The recognition of the value of 'unpaid' work is a major development in the area of APL and is particularly important to unwaged adults. Each year 50 per cent of the population undertakes some kind of unpaid work; 1.4 million people spend more than 20 hours a week caring for sick, disabled or elderly people, and approximately six million women of working age are not currently in paid employment.

'Paid' employment is actually only one type of 'work', and is only one out of a number of activities in most people's lives. The NCVQ identifies five main areas of significant activity that people engage in – family, employment, study, voluntary work and leisure – and recognizes that competences may be acquired in any or all of these. To place a greater or lesser value on competences from any one activity would be falling into the trap of equating work exclusively with conventional paid employment. All organizations are now being encouraged to recognize competence acquired in all areas of work, including that carried out in the home and family, in the community and the voluntary sector.

The gains that can be made from this wider definition of work, and thereby the enormous 'pot' from which competence can be gained, are to everyone's benefit. There are some groups, however, who stand to gain significantly from this development. Women have for years fought to have their skills and expertise as carers recognized, and demanded the status they should be accorded for the work they do. Recognition has been slow to come but, with the acceptance of unpaid work as an equally valid activity as paid work for the acquisition of competences, significant gains have been made.

What are the benefits of APL?

The benefits of APL can be identified for individuals, organizations and providers of education and training. Derek Crossland in the REPLAN Resources Pack on the *Assessment of Prior Learning and Achievement* lists these:

TO CLIENTS

For qualifications purposes, APL means that:
- qualifications or credit towards them can be given to clients who demonstrate they have reached the required standards;
- time required for completion of training schemes may be reduced;
- training scheme costs may be reduced;
- the unnecessary duplication of learning can be avoided; and
- it may be uncovered that the client is about to enter an inappropriate scheme, and changes can be made in time.

For gaining employment, APL can:
- produce a systematic overview of past learning and achievements;
- supply lists of verified competences;
- be used as evidence for interview purposes; and
- show personal characteristics.

For personal development purposes, APL can be used to:
- encourage self-assessment;
- gain a better understanding of oneself;
- increase self-esteem;
- increase self-confidence;
- increase motivation;
- provide a foundation for further development; and
- help maximize potential.

TO EMPLOYERS

For applicants, APL can:
- assist in making an informed judgement on clients' suitability for a particular job;
- match existing competences to job specifications;
- save time on assessing applicants;
- reduce the costs of recruitment; and
- match personal characteristics and attributes to a person specification.

For existing employees, APL can:
- reduce time out for training;
- recognize competences achieved at work;
- assist training needs analysis;
- enable easier progression;
- be linked to staff-appraisal schemes;
- build staff confidence;

- assist in the retention of staff;
- assist human resource management;
- help produce a training programme tailored to meet both the individual and organizational needs; and
- improve job satisfaction.

TO PROVIDERS OF EDUCATION AND TRAINING

APL can:
- enable them to reach new client groups;
- help them to tailor training to individual needs;
- assist them in making informed judgements on client needs;
- encourage more flexible and responsive delivery systems;
- maximize use of existing resources;
- avoid teaching clients things they already know; and
- improve quality of provision.

(Crossland, 1991)

Implications for adults

APL provides adults with an opportunity to have existing skills and expertise recognized, regardless of formal qualifications. It enables employers to set in motion practices that will get the most out of employees by recognizing skills and potential in a more comprehensive way than was previously done. It facilitates an organizational approach to vocational education and training that recognizes the value of work-based activities and builds on them.

APL has the potential to act as a catalyst for the development of a holistic approach to vocational education and training that could help alter the face of training in the UK.

New skills for the delivery of APL for adults

At present, colleges of further education are at the forefront of APL development because they have experienced a surge of support and enthusiasm for the process, and their project development has tried and tested some of the theory.

In some colleges, assessment units have been set up which offer APL to adults. In Birmingham a major project has just been completed in collaboration with colleges and voluntary groups on the assessment and accreditation of unpaid work.

Schemes for the unemployed are also offering some APL services, although difficulties remain over the limited resourcing available for the delivery of these programmes.

Like so many good ideas, however, and despite APL being an integral part of the NVQ process, delivery and access are patchy. All organizations which are trying to provide APL are struggling with the apparent cost implications, the different policies and financial regulations emerging from the awarding bodies, and very limited time to undertake the development work that is necessary to get the process off the ground.

The implementation of APL into vocational education and training structures has implications for both the way such education and training is delivered, and the skills of the staff concerned.

The traditional concept of starting a course in September and sitting the exam the following July is no longer relevant. As with NVQs, embedding APL into institutions and organizations offering education and training requires much more flexibility of delivery and a greater commitment to the needs of individual learners. Expert systems and computerized databases will ease the process, as will the design of comprehensive, coherent systems for recording and documenting activities. Modularization of further education provision, along with the development of new skills for those people involved in delivery, are vital if APL is going to work.

'Mentoring', 'counselling' and 'assessing' will have to become an essential part of the education and training provider's brief, and these skills will need to be accredited through the national standards framework of the Training & Development Industry Lead Body. The referral and networking activities of providers will also have to be developed to ensure that the skills available are put to the best use. A local 'map' of who can provide what and when on APL would ensure that there weren't too many new wheels re-invented, and that ideas and expertise were made available and shared.

The French 'bilans de competences' centres are staffed primarily by occupational psychologists, although this is not an approach that is being adopted in the UK.

What are the implications for the unemployed/unwaged?

Access to APL could open up new opportunities for unemployed people and enable them to acknowledge existing skills and potential in a more effective way than ever before. The identification of education and training needs which may have been hidden for years could also assist individuals along new routes, and into new opportunities.

Resourcing APL is proving to be a difficult and vexed issue, however. In the case of colleges of further education there is a lot of goodwill and commitment towards the concept, but a great deal of anxiety about the cost. Even the arguments about the long-term cost effectiveness of giving people detailed and accurate assessment and accreditation prior to entering education or training, and thereby decreasing wastage and drop-out rates, have not yet won through. Some of the barriers being identified are real; others are a symptom of a totally new concept not yet embedded into the infrastructures of education and training or industry.

For APL to be moved from the margins to the mainstream, collaborative strategies must be developed both within institutions, and between institutions and organizations involved in APL delivery. In this way resources, skills and expertise can be pooled, and the structures necessary for the effective delivery of APL can be established across 'areas'. Marketing would be comprehensive, and acknowledge the different partners and the part they have to play.

Consortia arrangements between employers, colleges and through the work of TECs could move this whole area on considerably. The development of an APL 'map' across TEC areas would be invaluable, although this would need to be underpinned by an agreed policy of access for all – employed and unemployed – to the services on offer, and TECs would need to consider very seriously the issue of vouchers for APL for unemployed people.

There is a way forward with APL, but it will take innovation, collaboration and commitment.

5. Developing competence-based learning through NVQs

What and why

> NVQs are a certificate of the *competence* of people in some area of work, rather than just their academic knowledge.
>
> This means that they will testify to what people *can do* (which encompasses knowledge and skills, and the ability to apply them in practice in the real world of work), rather than just what they *know*. (NALGO, 1990)

The National Council for Vocational Qualifications (NCVQ) was set up by the Government in 1986 with a brief to establish a coherent national framework for vocational qualifications in England, Wales and Northern Ireland, and improve vocational qualifications by basing them on nationally agreed standards of competence required in employment.

NCVQ is not an examining or awarding body. The existing examining and awarding bodies, such as BTEC, C&G, RSA, PEI, etc, submit qualifications to NCVQ for validation according to strict

NCVQ criteria. It is these criteria which underpin the rationalization process, ensure relevance of all qualifications to the needs of the work-place, lay down clear guidelines for an 'outcomes' competence-based model and make clear statements about opening up access routes to qualifications, and equal opportunity procedures.

The fundamental criteria

The fundamental criteria laid down by NCVQ for the validation of awards as national vocational qualifications are that they must be:

- based on national standards required for performance in employment (laid down by Industry Lead Bodies), and take proper account of future needs with particular regard to technology, markets and employment patterns;
- based on assessments of the outcomes of learning, arrived at independently of any particular mode, duration or location of learning;
- awarded on the basis of valid and reliable assessments made in such a way as to ensure that performance to the national standard can be achieved at work;
- free from barriers which restrict access and progression, and available to all those who are able to reach the required standard by whatever means;
- free from overt or covert discriminatory practices with regard to gender, age, race or creed and designed to pay due regard to the special needs of individuals.

A model based on national standards

The development of standards was first identified as an important objective for vocational education and training in *A New Training Initiative* (MSC, 1981). Standards are based on the needs of employment and embody the skills, knowledge and level of performance relevant to the work activity.

Each NVQ is made of a number of units in which the national standards for a particular occupation are broken down into competence statements. These competences are set down by Industry Lead Bodies (ILBs) which are made up of representatives from the occupational sector trade unions and other specialists in the field of education and training.

Units can be taken separately and evidence accumulated until a full NVQ is achieved. Completion of an NVQ, or any individual unit, will not be dependent on attendance at a particular course, or even through a predetermined scheme or series of activities. It is anticipated that the bulk of assessment of NVQs will take place in the work-place, or in a situation that is as close to a real work-place environment as is possible.

The traditional concept of initial training taking place prior to commencing work, 'bolt-on' work experience, 'paper-base' qualifications with final examinations, and the notion of 'time served' no longer applies. NVQs are a much more flexible approach to the development and acquisition of work-based skills than traditional qualifications, and have the potential to facilitate access to vocational qualifications for a much greater range and number of people, particularly adults.

Central to the NVQ process is the development of personal records of achievement in which individuals can collect evidence of competence as and when it is acquired, either this be as part of an existing development plan, or more longer term goals.

Also vital to the whole process is the integration of work-based activity with the acquisition of underpinning knowledge, and the equal access of all to vocational qualifications.

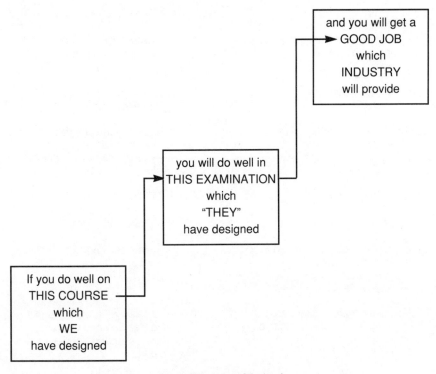

Figure 2 *'The old view'*

PART II: PROVISION FOR UNWAGED ADULTS

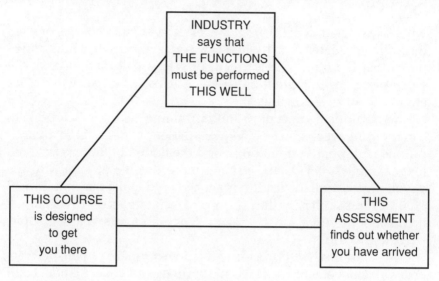

Figure 3 *'The new view'*

COMPETENCE BASED

1. Ability to do job effectively

2. Time not a factor

3. Criterion of success is ability to do the job

4. Entrance requirements not of primary concern

5. Flexible scheduling of learning

6. No fixed rules as to how, when or where learning accomplished

7. Opportunities provided to acquire competences in practical field or on-the-job experiences

8. Learning (competences) presented in small units or modules

TRADITIONAL

1. Knowledge of subject and practical tasks

2. Specified time limits

3. Criteria of success are examination grades

4. Entrance requirements are important; previous qualifications often required

5. Instruction scheduled into blocks of time. Academic year and infrequent mass registration

6. Classroom teaching most common approach to instruction

7. Practical field experiences being limited

8. Learning (subject matter) organized into courses representing academic time units

Figure 4 *Comparison of competence-based and traditional schemes*

SIX KEY ASPECTS IN DELIVERY

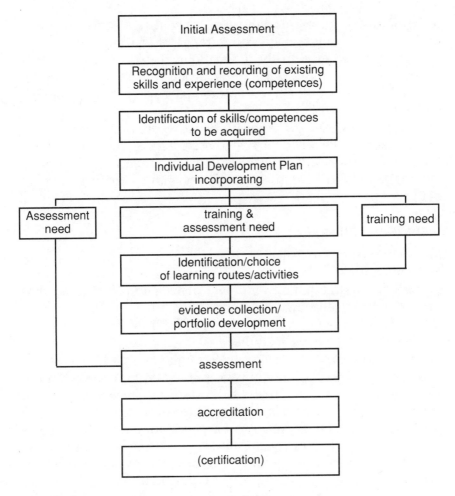

Figure 5 *The NVQ process - a model*

Development implications

While the standards-based approach to education and training has its roots in policies dating from as far back as 1981 – and very clear progressive, staged development can be identified over the last ten years, not least being the development of competence-led programmes in the YTS – the whole system is still struggling to make the profound impact on the UK education and training system that was expected and, indeed, is needed.

The principles of the NVQ framework point to a system of learning and achievement which demands from all providers – work-based or otherwise – greater flexibility; learner-centred approaches;

recognition of prior learning and experience; assessment on demand in a variety of contexts; access to a range of learning options and systems; greater support for 'learners' in the work-place; greater investment in 'training'; clear progression routes to other opportunities; and institutional/organizational approaches to training and development.

Whether NVQs are delivered in colleges or in companies, or even in a partnership between the two, there are enormous implications for the infrastructure of all organizations as they currently exist, and also for the skills base of existing staff who may be involved in NVQ delivery.

Implications for adults

Motivating people to learn within a coherent framework of education and training, creating an individual focus through personal profiles and individual development plans, and offering transferable skills and relevant work-based qualifications are all fundamental to the successful delivery of NVQs. How relevant is this framework to the needs of adults?

Firstly, unlike previous vocational training, which frequently required formal pre-entry qualifications, the NVQ framework is designed to be open access. Previous educational achievements are not deemed to be the only route to vocational success, and are not prerequisites for the development and accreditation of a person's work-related skills. Nor is age a factor. NVQs can be gained at any time in a person's working life – a significant move away from the principle of an initial training period followed by a period of work, with top-up skills being provided according to the needs of the employer. In this respect, the implementation of NVQs has the potential to change the nature of training in the UK.

Secondly, NVQs are awarded on evidence of actual work-related performance, rather than theoretical knowledge. Statements of competence demonstrate that the holder has the ability to perform a range of work activities and possesses the skills, knowledge and understanding which underpin such performance. In this respect adults can be recognized for existing skills through a framework of unit-based competence statements, and thereby have much greater access to accreditation than ever before. The assessment and accreditation of prior learning (APL) is an integral part of NVQ delivery.

Thirdly, the acquisition of skills and knowledge through defined competence statements can be achieved flexibly, and at the person's own pace. There is no time limit set for accumulating units leading to an NVQ, nor will one particular mode of learning be stipulated.

It may well be that individuals will find a mixture of on-the-job training, simulated exercises, distance learning and some college courses appropriate. Adults will be able to choose the best route to assessment and accreditation according to their own needs and circumstances. The National Council is also working on 'generic competences' which may be common to several different occupations or work roles, and which, when defined, will facilitate the credit transfer process.

Lastly, NVQs offer opportunities to develop transferable skills which can be accumulated and used as the basis for credit transfer between different qualifications. The existence of profiles and national records of achievement will ensure permanent records of assessed competences are kept, and will enable adults to develop and build on their skills and abilities over any given period.

How can the adult unemployed/unwaged benefit?

The development of a framework of national vocational qualifications is a significant step towards providing adults with greater access to vocational education and training, particularly as it acknowledges prior achievement and aims to accredit this.

How this opportunity is applied to the adult unemployed is a cause for some concern. NVQs are largely work-based and relate to work-based skills and activities. Accreditation can only be achieved in a work environment, or the nearest thing possible to this. All the Government targets for NVQs refer specifically to the 'employed', thereby marginalizing yet again those members of the work-force who are without 'paid' employment.

Clearly, the adult unemployed will have a whole range of skills and experience to bring to the process of assessment and accreditation for NVQs but, being out of the work environment, they will not be able to access the system in the same way as the employed.

Government programmes for the unemployed are, in some cases, incorporating NVQs, but, as stated earlier, resourcing for these programmes is limited, and the development of systems and access routes for unemployed people to quality education and training and the achievement of NVQs is often beyond the means of Programme Managers.

Some further education colleges are also going down the road of offering access to APL services to adults whether they are employed or not. They too, however, are struggling with the cost implications of a resource-intensive activity in an environment which is emphasizing provision that will 'generate a surplus'.

TECs have the responsibility for Government programmes for the adult unemployed. The TECs will need to think very creatively about systems and access routes to NVQs for unemployed adults, or risk the development of a whole 'under class' of unemployed people to whom access to the potential offered by NVQs will be denied.

If the framework of national vocational qualifications is going to make the major contribution to raising the skills base of the UK that it is hoped, then urgent steps need to be taken to ensure that the unemployed can get equal benefits from the system, and thereby equip themselves to re-enter the labour market.

6. Developing and embedding core skills

> The limited opportunities which most adults have to take part in education and training would be most effectively used if all curricula were designed to include
> - the processes of negotiation and needs of analysis, counselling, guidance and assessment, and the accreditation of experience and consideration of relevance, and
> - a core of skills which forms a basis for learning and includes the areas of communication, literacy, technology, creativity, and planned learning strategies
>
> within programmes responding to learning needs arising in the context of specific adult life-roles. (FEU, 1987)

Research and development on core skills has recently been undertaken by the National Curriculum Council, the Schools Examination and Assessment Council, the Further Education Unit, the CBI, and the National Council for Vocational Qualifications. While much of this work is fairly advanced it is nevertheless difficult to predict the final outcome.

What we are concerned with here, however, is the place of core skills in adult vocational education and training, and, indeed, whether current developments and thinking are taking account of adult needs.

Developments in the 1980s

During the 1980s a considerable amount of work was undertaken on core skills. Among others, Margaret Levy in *The Core Skills Project and Work Based Learning* (1982–1985) looked specifically at the core skills of young people on YTS schemes, and the Further Education Unit identified the need for a coherent national strategy for adult vocational education and training, enshrining the right of all adults to a core of basic skills and competences.

In *Supporting Adult Learning*, published in 1987, the FEU laid out a possible framework for a core skills curriculum giving adults 'opportunities to acquire a core of skills and competences, to form the basis for the more specialized demands of learning related to adult roles'. While identifying a set of core skills for adults, and undertaking this in a context of a minimum right to basic skills and competences for adults, the FEU also recognized the need for core skills to be offered in all school and pre-vocational curricula.

Some people may acquire key competences as part of their early education, but many will not. Age cannot be an inhibiting factor in the acquisition of a set of vital core skills, and ongoing access to opportunities to acquire such competences was a key aspect of the proposal. Indeed, a recognition of this need was one of the reasons the research was carried out in the first place.

The list of core skills recommended makes interesting reading – communication, literacy, numeracy, technology, creativity, learning skills and role-related learning – and was drawn up on the basis of adults having the right to basic skills which would enhance all aspects of their life.

The main platform for these proposals was much the same as now – a background of social, economic and technological change – but there was an element of social justice and equality not apparent in current thinking:

> We suggest that an outcome of all (curricular) provision should be enhancement on the part of learners of:
> - adaptability and flexibility – the readiness and understanding required to transfer skills by applying them appropriately in contexts other than those in which they were originally acquired, and to seek further knowledge and skills independently and from new sources;
> - assumption of responsibility and the exercise of initiative;
> - understanding and acceptance of other individuals and groups leading to the successful exercise of interpersonal skills;
> - intellectual skills, including specific study and learning skills, and decision-making and problem-solving skills;
> - confidence and motivation.
> (FEU, 1987)

No serious attempts appear to have been made at this stage, however, to produce a comprehensive set of core skills relating to all vocational education and training, and, indeed, academic qualifications.

In 1989, Kenneth Baker, then Secretary of State for Education, addressed the National Association of Teachers in Further and Higher Education (NATFHE) and stressed the need for future employees to have broader skills, knowledge, understanding and greater flexibility. Economic need was driving the debate, and Baker later asked the National Curriculum Council and the Schools Examinations and Assessment Council (SEAC) to look at 'core skills' for 16- to 19-year-olds. Several reasons can be identified for this move at that time:

- the need for greater uniformity and continuity between A levels and vocational training;
- the need to identify, within the NCVQ framework, a set of core competences which could be acquired and transferred between occupational sectors;
- the need to develop a more flexible work-force by promoting adaptability and mobility within and between jobs and occupations; and
- the need to provide greater 'breadth' within NVQs, which were starting to be criticized for their over narrow vocational focus.

There has been some dispute between the various bodies involved about the nature of core skills and what they should actually be. Core units are, however, being developed in two phases by the NCVQ, the body which appears to be taking the lead in this area at the moment. The first covers communication, personal skills and problem solving, and the second includes numeracy, information technology and modern languages.

Discussion has also taken place about the context in which such skills acquisition should take place and areas such as economic and industrial understanding, education for citizenship, career education and guidance and health education have been talked about.

A comparison

While the work carried out in 1987 by the FEU and that currently being undertaken by the NCVQ both aim to identify a set of core competences in vocational education and training, there are one or two significant differences: the FEU work was identifying a better way forward specifically for the education and training of adults, while the NCVQ's work on the core is driven by the need for greater transferability and 'breadth' in its qualifications. The FEU was looking at the issues from a standpoint of equality and the difficulties of access to education and training that many adults experience. The NCVQ is responding to central Government requirements as well as pressure from external bodies unhappy with existing NVQs. The FEU work calls for a coherent national strategy for the vocational education and training of adults; the NCVQ makes no such demands, and, in fact, does not even focus on the particular needs of adults in its current work on core competences.

A core for adult vocational education and training

Much of the work being undertaken on core skills could, and indeed will, also benefit adults, especially those who have access to NVQ accreditation.

To ensure, however, that all adults have equal access to opportunities to develop, upgrade or even acquire for the first time, a set of core skills and competences there will need to be an underpinning commitment to a minimum entitlement and right of access to core skills in adult vocational education and training. At present there is no such commitment, no such core and certainly no clear access for adults to develop and upgrade skills to an agreed minimum set of competences and skills.

Summary

Far from being developed and enhanced, education for adults that is not strictly work-related is being squeezed and marginalized more than ever as a result of recent legislation, and the removal of colleges of further education from local authority control will create an even more piecemeal and disparate service, one that is difficult to locate, frequently difficult to access and has an emphasis on revenue-generating provision.

Entitlement to a minimum defined 'core' for adults – whether employed or unemployed – has been on the agenda of adult educators for many years. It would increase and improve the skills of vast numbers of adults, and go a long way to genuinely upgrading, and raising the skills level of, the UK work-force in the way industry is increasingly demanding. It would enable unemployed adults to be considered on an equal footing with those in paid work, and help to create a 'learning culture'. Why it has not been recognized and implemented at central Government level remains a mystery.

6 Measures to Facilitate Changes in the Current System

Systems maintenance is essential – it keeps the ship afloat, never mind where it is going. (Thurley, 1990)

So far in the book, attention has been drawn to the specific needs of the unwaged in relation to vocational education and training. Who makes up this particular group of people has been described, along with why, in many ways, what has been, and still is, on offer for them is inappropriate to them as individual learners. Why some adults have a negative attitude to learning as well as some of the personal and institutional barriers they face on returning to learning and training have been explained, and a belief that there is a better way of approaching training for them has been expressed.

Although the theory of valuing individual learners and creating a lifelong learning culture in which they may flourish is idealistic, it is nevertheless possible to transfer its messages into practice not only to the advantage of the individual learner but also to fulfil the stated needs of those who are concerned with this country's growth. Therefore, an attempt has been made to link its development as an integral part of vocational education and training with the development of the economy and the needs of the business community. To succeed in the future, the UK needs to ensure that all members of its adult work-force are well trained and educated in ways which will encourage them to want to continue to learn throughout their lives. This country must become like other countries, in which education and training are seen as a gateway to success, independence and opportunity.

At the moment there is an apparent shift along the lines described for those who are already in work, but this new approach must be advocated not only for them but also for the unwaged so that they can contribute to the growth of the economy and grow as individuals.

This cannot be achieved overnight, or by a single person or institution. So, what follows are a number of measures which need to be taken by the Government, by institutions and organizations, by tutors, and by individuals in order to change the system in which vocational education and training is delivered and the individual's attitude to lifelong learning so that, in the long term, the prospects for the country and its people are greatly improved.

The Government

There are several things that the Government must address if the suggestion of marrying the economic approach and the humanistic approach to vocational training for unwaged adults is to be realized.

Financial commitment

First of all, the Government must make a real financial commitment to learning and training opportunities for the groups of people who are disadvantaged and, therefore, often overlooked when an investment in training is being considered. They must abandon the current practice of offering short-term opportunities and create much longer, better supported situations in which people have the chance to develop something more than a skill which is soon out of date. Close attention should be paid to Sir Christopher Ball's 'vicious circles' which he describes in the RSA/Industry Matters report *More Means Different* (1990). He relates low investment with low standards of education and training, low skills, low productivity and profits, low salaries and wages, low satisfaction and low aspirations. What would be more desirable, he argues, is a 'virtuous circle' for the 21st century in which high investment would mean high standards of education and training, high skills, high productivity and profits, high salaries and wages, high satisfaction and high aspirations. This would create a society in which 'an appetite for learning is as instinctive as for eating or drinking' (Tuckett, 1991).

In *Towards a Learning Workforce*, Alan Tuckett (1991), the Director of NIACE, says that: 'Fostering such an appetite and satisfying it would produce an improved economic performance for Britain. But the merits in learning are more than narrowly instrumental. A society committed to active citizenship and a developed democracy benefits disproportionately from an informed, confident and skilled

PART II: PROVISION FOR UNWAGED ADULTS

adult population. People with good learning experiences seek more for themselves, their families and their colleagues.'

This, surely, is the biggest challenge facing people who firmly believe in this way of thinking, and it is fundamental to what they hope to achieve. There are no easy answers when this aim does not seem to be shared by central Government, so these beliefs must be continually stated openly and, where possible, tutors must lobby for the appropriate legislative and financial changes to be made. Otherwise, all other initiatives, from the grass roots upwards, will be difficult to sustain.

Using legislation

As well as making a substantial financial commitment to the learning culture, a Government is needed which believes in this philosophy to the extent of actually using its legislative powers to establish an adult entitlement to lifelong learning and training. By using public funds to create such an entitlement, unwaged adults, who may have missed out in terms of compulsory education, would be given that all important second chance.

What is meant by entitlement, in this instance, is some kind of promise by Government which would guarantee every adult periods of supported education and training throughout their lives. This is a view shared by many people now and there are various interpretations of exactly what this entitlement should be. For example, David Miliband, in *Learning by Right*, published by the Institute for Public Policy Research (1990), suggests that every adult should be offered three years of post-school education and training throughout their working lives. As a start towards this goal, he suggests a national target of five days' education and training per adult per year. This entitlement would be open to all adults not in full-time education and aged between 16 and 65. He says that educational leave could be taken part-time, or accumulated over a number of years. Policies for such leave are already in place in many European countries and a commitment to lifelong learning is enshrined in the European Community's Social Charter. Of course all of this would only benefit those in work at the moment, but the principle is there and, once in place, it could easily be transferred to the unwaged and replace the current unsatisfactory shorter term schemes. Such an entitlement would move a long way towards improving the poverty of aspiration so often referred to in relation to the adult population of this country, because it would be a demonstrable indication to working people that they, as individuals, were valued.

A commitment to such a scheme, or an adapted version of it, has enormous implications for every kind of provider of vocational education and training for adults, in particular the TECs and colleges of further education. This is because large numbers of appropriate courses would need to be made available and, therefore, the whole infrastructure would need careful restructuring and monitoring for quality.

Again this is another aim which tutors may wonder how to achieve, and what relevance it has to their day-to-day work. But it must be seen that it makes sense to have such a philosophy locked into our legislative system. Tutors might also like to question why Britain was the only member state not to sign the European Community Charter which declared that 'every worker of the European Community must have access to vocational training and to receive such training throughout his/her working life'.

Finally, in terms of legislation, and with particular reference to those unemployed adults who may wish to participate in some form of vocational education and training, the Government needs to move away from the current practice of allowing unemployed people to study for only 21 hours a week. Adults should be positively encouraged to study if they want to and not demotivated by placing restrictions on them.

Similarly, those who choose to study part-time and work part-time are discriminated against because they are not eligible for full-time grants or employer sponsorship. Legislative changes which make it easier to participate in vocational education and training would be an enormous step in the development of a lifelong learning culture.

Offer leadership

One of the key factors, identified in this book, which prevents the real development of the vocational education and training of unwaged adults is the attitude which they themselves have towards learning. This resistance to learning is not entirely the fault of the individual, but is due to a number of external factors such as peer-group pressure, poor experience of school, adverse parental attitudes, lack of confidence and low self-esteem. Also, the system of education in this country tends to favour high achievers and there is a great deal of emphasis placed on the achievements of this elite group rather than everyone in the system.

We believe that the Government has a responsibility, and the ability, to create a learning culture and develop a much more positive attitude to learning. Not only could it adopt the appropriate rhetoric but it could also draw up clear, staged plans which would gradually

PART II: PROVISION FOR UNWAGED ADULTS

enable a better system to operate, to the benefit of the whole of society rather than just an elite few.

Finally, in terms of the Government's role, NIACE's proposal that there should be a single ministry for education and training should be wholly endorsed. The current system, with separate departments, often results in a lack of overall coherence. This would then give appropriate centrality to issues affecting adult learners.

Institutions and organizations

The role of those institutions and organizations which have a responsibility for providing vocational educational and training opportunities for the unwaged – ie, colleges, TECs and private training providers – is extremely important in developing a lifelong learning culture.

First of all, in general terms, they need to consider the points made earlier about appropriate methods of promoting, delivering and assessing the work they do with unwaged adults and work towards their full implementation. This will have the effect of increasing access and generally making the whole learning experience more attractive so that adults will want to return to it again and again.

Secondly, they need to work together to develop co-ordinated policies and programmes which allow people to progress through the system at their own pace and in ways which are relevant to their own personal circumstances. The partnership approach is not a new idea, but is still applicable when working towards securing adequate provision for the unwaged as well as developing a learning culture.

Thirdly, the role of the TECs must be seen as key to the development of the learning culture which has been described. They have both the interests of the local community and the local employers to satisfy. Therefore they are in an ideal position to ensure that high standards of training and education are maintained, that there is appropriate provision and adequate guidance and advice services and that the staff employed by their trained providers are fully aware of the needs of their adult clients. From their strategic positions they can inform the Government about the kinds of resources needed to create the well-trained and flexible work-force which will ultimately help to improve the economy of this country. Hopefully, this will eventually influence the Government to change its current policy and practice. They are also in an ideal position to create the kinds of partnerships described earlier between colleges, private trainers, the voluntary sector, employers and adult education which are essential for student progression and individual development.

Finally, although they cannot be described directly as providers, the importance of employers in terms of the development of our learning culture must not be forgotten because they do have a vested interest. They want to see economic growth and, therefore, need to acknowledge that a well trained and flexible work-force is the answer. Many large companies, some of which are illustrated in the case-studies, are already committed to developing a better educated work-force. While it is appreciated that it is not so easy for smaller companies to organize massive in-house training programmes, a collaborative approach could be adopted which would be of benefit. Again, this is not something which will directly benefit those unwaged adults who are seeking some form of education and training and, ultimately, employment. However, what this approach by employers will achieve is a general shift of attitude towards learning which will eventually influence the whole population.

The kind of Government intervention which was described earlier in this chapter involves a commitment to long-term aims which must continually be strived for. But the more localized responses by colleges, TECs and training providers which have just been referred to are much more achievable and can in many instances be established in spite of current legislation and inadequate funding.

Individual tutors

Individual tutors must never underestimate the importance of their role in the development of a learning culture within vocational education and training. They are the ones who actually provide a service for those unwaged adults who have not been well served in the past. They are the ones who understand them, relate to them and understand their needs. Therefore, what they need to do, as well as continue with the good work which they have already started, is:

- ensure that all their colleagues who are working with the unwaged take full advantage of the training opportunities which are available to them so that they fully understand this particular client group, the methods most appropriate to teaching them and the kind of curriculum which is best suited to developing their full potential;
- make what they do as interesting, exciting and relevant to the lives of unwaged adults as possible so that learning becomes a

positive experience to which they will want to return again and again;
- find time to publicize what they do and its successes so that ordinary people, who would normally think that education and training was not for them, are attracted to what they provide;
- use the press and the media to 'educate' people about the new developments such as accreditation of prior learning, NVQs, open and distance learning, credit accumulation and transfer, and modular learning methods, so that they begin to understand that the world of education and training has changed and is much more in tune with their lives and needs;
- advertise their childcare facilities, their access for people with disabilities, their language support for bilingual learners and other learning support services which help to overcome barriers to adult participation in study;
- get themselves, and what they do, out there where the people are, in community centres for example, to show them that learning does not have to take place in a formal institution which may be unfriendly and unfamiliar;
- try to make the environment in which learning and training takes place as attractive and welcoming as possible so that the adults who come through the doors do not feel threatened or anxious;
- join or develop networks of contacts from a wide range of organizations so that they can assist in terms of promoting provision, providing progression routes and generally keeping staff up to date with local, regional and national developments;
- continue to lobby for additional resources, better facilities and a more 'mainstream' attitude and approach for what they do;
- find out about schemes which demonstrate good practice in terms of their work with unwaged adults and use them as examples when making requests for their own service; and
- create a good adult guidance service where people can come and discuss suitable courses and ways forward so that they can make informed choices about what they do.

Individuals

It is not easy to find a role for the adults being discussed in this book in terms of helping to develop a different approach to their education and training. However, as adults, they surely have a right to understand what is happening and why. Once they are aware and are participating in some form of provision, then, if it has been successful, they also could put pressure on the Government and organizations to increase resources and adapt more to their needs.

In an ideal world, unwaged adults who wanted to return to education and training would be able to do so at no cost to themselves. Therefore, free provision and a reasonable living allowance should be advocated as a right. However, an ideal world does not exist and, until it does, the possibility of some adults contributing financially and investing in their own learning to promote a learning society must not be ruled out. Tax relief and low cost loans could be made available to those in a position to make their own contribution.

Finally, it is important to say that no single person or organization can transform overnight the way in which vocational education and training is developed to meet the needs of our changing economy. However, it is hoped that the combination of ideas, good practice and proposals put forward in this book will go some way towards creating the necessary changes and stimulate a new attitude and approach to education and training in this country.

Summary

What Part II has highlighted is the value of a number of approaches which not only encourage adults to return to learning by removing barriers to access, but also help to stimulate a lifelong interest in learning. The emphasis again must be on duality of purpose – the role and place of the individual adult learner should be given the same prominence as the immediate needs of the economy. Only in this way will individuals realize their full potential and industry benefit from a skilled, mobile work-force.

Certain approaches contribute to achieving this duality and these have been documented here: guidance, learning and teaching methods suited to adult needs; systems of financial support for adult learners; APL; competence-based qualifications (NVQs) and the commitment to a central core entitlement for adults.

These key attributes of a successful and effective infrastructure for adult vocational education and training need to be more widely developed and implemented and a commitment made to the realisation of individual adult potential through this kind of good practice. Unwaged adults in particular will benefit from these changes.

Policy and practice must be developed by the Government, institutions and organizations, tutors and trainers and individuals to ensure that an appropriate education and training infrastructure for unwaged adults and, indeed, all adults, exists.

PART III

Responding to the Challenge

CASE-STUDIES OF GOOD PRACTICES IN ADULT VOCATIONAL
EDUCATION AND TRAINING

Introduction

So far there has been a great deal of discussion about the need to change existing practice in vocational education and training for the unemployed and the unwaged, so that a different learning culture emerges which will benefit both the individual learner and the economy of this country. A lot of rhetoric about what should and could be done and quite a bit of criticism of the existing state of affairs has been provided. Good practice and the potential for the unemployed or unwaged learner in the present economic climate has been examined. The argument for acknowledging and encouraging individual learners' needs has been developed, and the possibilities which could emerge for the individual learner if the curriculum were broadened have been looked at.

Now it is time to provide some concrete evidence in support of this argument for change.

For the past 12 years, the authors have been involved in the field of adult education and training. During that time they have seen and heard about many interesting, innovative and practical things being done with the adult unemployed, the unwaged and adults in general, both in this country and in Europe – projects, courses and programmes which have demonstrated a real commitment to putting the needs of the individual learner at the forefront of planning and delivery. They have seen how effective this approach has been in developing people's self-confidence and setting them on a lifelong path of learning and training.

PART III: RESPONDING TO THE CHALLENGE

The case-studies which follow have been specially selected from their experiences and contacts because they truly reflect their philosophy and demonstrate the kind of good practice which they would advocate. Also, and this is very important in terms of what has been said so far, in most cases they have been extremely successful in attracting to education and training those adults who, traditionally, have been reluctant to return to study or participate in any form of further learning or training. Two case-studies have also been included (the Tourism Innovation Project and the Euro-Consultants Course) which are aimed at unemployed people with a background of higher education. These have been included because they both reflect the same philosophy and demonstration of good practice and both are transnational. It is hoped that this kind of transnational initiative could later be developed to cover a much wider target group.

Some of the selected examples describe training schemes which are for people in employment. These are featured because they have demonstrated an ongoing commitment to putting the needs of the individual first rather than those of the institution or organization. This and the high level of success in encouraging an ongoing interest in learning for those concerned are most impressive. Therefore they have a transferable message in terms of the client group, which is the main focus of this book.

Although the schemes or programmes which have been included are different in many ways in terms of sources of funding and providers, there is actually a common thread to all of them. This is the emphasis they all place on putting the needs of the individual first, and this is both central to the argument for change and key to the process.

Other important factors which emerge from the case-studies are that they:

- consider the needs of the unwaged adult;
- acknowledge the personal and institutional barriers which unwaged adults face when they return to learning;
- use flexible styles of delivery which are appropriate to unwaged adults;
- use specially designed courses for this particular group;
- have committed, well trained staff;
- provide guidance and advice to their students;
- demonstrate collaboration;
- acknowledge short- and long-term market needs; and

- recognize a duality of purpose between the needs of the individual and labour market.

They all demonstrate a high success rate in attracting back into education and training those adults who have traditionally not participated. Since their positive results show that this learner-centred approach does work, questions need to be asked about why the schemes described seem to be:

- on the margins rather than in mainstream provision;
- dependent on short-term funding/projects;
- dependent, for their continuation, on job-related outcomes;
- too short;
- underfunded; and
- vulnerable to cut-backs.

In spite of this, the work being done in the case-studies is very encouraging because it strongly reinforces what has been said throughout this book. Also, if so much can be achieved in a climate of uncertainty about continuity, with low funding and the marginal nature of the provision, just imagine what could be done, in terms of developing a lifelong learning culture in this country in which vocational education and training had an integral place, if there were a *real* investment in these ideas.

7 An Adult Route to Qualifications

CONTRIBUTOR: LOUISE ROWE, NORTH LINCOLNSHIRE
COLLEGE OF FE

Accreditation of prior learning (APL) is one of the most exciting developments in recent years. Although still in the early stages, its arrival and acceptance by the world of education and training will have an enormous impact on future opportunities for unwaged adults and will eventually revolutionize the qualifications base of the UK work-force.

Access to a wide range of courses will be made so much easier if informal learning, voluntary employment and voluntary experiences can be credited and recognized in place of more formal qualifications. In this way, those who have been traditionally unrepresented or under-represented in education will get that all-important second chance.

At North Lincolnshire College we wanted to develop a centralized, regional APL service which would enable us to offer something which FE had not been able to offer before, ie, a flexible entry system. As a result of our APL project, there has been a radical change in our approach to assessment and qualifications because now we can give credit to what people already know and have done, rather than insisting that they should always attend a course first. We feel that our project is really exciting and important because it recognizes individual adults' achievements and then relates this to recognized qualifications. Our approach is, therefore, positive and based on what people can, rather than cannot, do.

What follows is a detailed description of our APL project which was originally funded by TEED. Responsibility for funding the project has now been taken over by the Lincolnshire TEC.

When the APL project first started, our main original objectives were:

- to develop a centralized regional APL service with both the Gateway Centre in Lincoln and the Gateway Centre in Peterborough;
- to establish a central management group linked to the TECs and the Adult Guidance and Careers Services in Lincolnshire and Cambridgeshire which would integrate and co-ordinate the provision of a coherent regional APL service; and
- to build on the foundations of the initial pilot project in APL and to provide further information for a national model.

In some quarters, reservations have been expressed about the desirability and feasibility of APL because it calls many accepted practices into question. However, we have been fortunate at our college because the people working there have seen that to adopt changes which increase the availability of opportunities to a much greater proportion of society must be a good idea. So, what we have aimed to do is deliver an APL service for individuals, groups and employers which identifies competence and matches those competences to units of NVQs. If competence can be proved, it can lead to the award of units or whole qualifications by any of the awarding bodies.

At first our APL service was limited to catering, construction, electronics, motor vehicle mechanics and office practice. This was later extended to include agriculture, horticulture, retail, hairdressing and beauty therapy.

To offer APL as we wanted to, on a regional basis, we had to ensure that the Gateway Centres at Lincoln and Peterborough were supported, across the region, by a network of access points. This meant that APL would then be within reach of a large section of the population and would not be just a college-based service.

So, bearing all of that in mind, we then offered the following service at different levels.

- *APL for individuals:* This initially meant that recruitment was carried out on an individual basis. Candidates were referred through publicity, then given appointments at the APL offices.
- *APL for groups:* Both mixed and single vocational areas were targeted for this service which was offered to employers, ET managers, voluntary organizations and HM Forces.
- *APL for employers:* Employers were offered APL audits to help structure and plan staff training

PART III: RESPONDING TO THE CHALLENGE

Access point staff worked within the above groups on portfolio production and then referred either groups or individuals for assessment and accreditation at the Gateway Centres. It was important to ensure that, if such a service was to be effective, it was flexible enough to meet the needs of all these clients. Figure 6 shows the actual process people went through as part of our service.

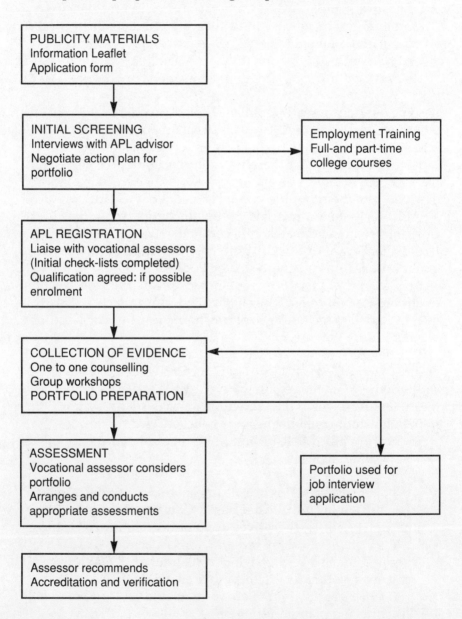

Figure 6 *APL process*

Outcomes of the project

In phase one of our project, the administrative structure was completed. Advisers and assessors had been trained and we had begun marketing the service to individuals and employers. By the end of phase one of the project in March 1991, 655 people had made enquiries at the two Gateway Centres and 160 enquiries had been made at the access points. Of the total 815 enquiries, 499 had been seen for the first interviews and 33 were either assessed/accredited or waiting for verification of their assessment. 114 were in the process of completing portfolios for assessment. The majority of candidates were registered in groups from employers. This was all achieved in 17 months.

Key aspects of the project

Phased implementation
It was important to allow a period of six months in phase one for preparation. This allowed time for good links to be made with access points, staff training and recruitment, the setting up of effective administration systems and the promotion of the service to individuals and employers.

Role clarification
In our project, the APL advisers who were appointed were generalists, not specific vocational advisers. Their role was to facilitate and guide the APL candidates through the process of portfolio production and final assessment – ie, from the initial interview to accreditation and follow up guidance. The advisers liaised with vocational assessors throughout the whole process. The role of the vocational assessors was to prepare check-lists for use by the APL advisers. These check-lists were formulated from units and performance criteria of the NVQs available in each vocational area. The assessors' key function was to evaluate evidence to determine whether or not the candidate met the standards required for accreditation and then liaise with the adviser in order to give candidates the best possible feedback on their options. Once the candidate had completed the portfolio of evidence, the assessors' task was to conduct an appropriate assessment and make recommendations for accreditation and, if necessary, top-up training.

Promotion of the service
This was very important and was carried out on a regional basis. It included visits to employers and training agencies, presentations to

groups of employers, training agencies, careers and employment services staff, further, higher and adult education staff, leaflets, posters and media coverage.

Adequate administration support
This was essential since it allowed the staff involved in the project to get on with the specific task.

Jobs and further education for trainees/clients

There is an obvious link between qualifications and employment in terms of APL, but it can also be used as a valuable tool to update and recognize existing competence. This links in with future prospects of promotion, etc. An additional benefit, less easy for us to quantify, but demonstrated throughout the project, is the boost to self-confidence and motivation which resulted from recording competence.

Guidance throughout the APL process can be intrinsically worth while. It helps to break down barriers to access to educational opportunities for adults.

APL can also be used by employers as a means of identifying competence as well as to help pin-point training needs.

There are clear, mutual benefits for individuals and employers who take advantage of APL. It will:

- facilitate training plans which can be tailored to meet specific needs;
- assist human resource management;
- give full recognition to competence gained at work, from general life experiences or independent study;
- reduce time out to attend courses;
- reduce training costs;
- assist in recruitment and retention of staff; and
- make staff appraisal and quality assurance more effective.

Assessment and recruitment

Assessment and recruitment of APL candidates on our project is done on a flexible basis and is not tied in with traditional course provision. In line with NVQ policy, there is no requirement that candidates should have acquired competence by specific methods or over a specific period of time. Clients have to produce a 'portfolio' of evidence and this expands assessment to cover all paid and unpaid areas of work. Inherent in this is the recognition that learning takes place throughout life and is not restricted to a school or training institution. The most important things about APL assessment are the ways in which competence can be translated into acceptable evidence and that it takes into account reliability, authenticity and recency of experience.

The assessment process is rigorous but not rigid and, where appropriate, a candidate may be assessed by demonstrating current competence in a work setting.

Feedback and evaluation

During the life of our project, continuous monitoring and evaluation took place. Reports were submitted to the Training Agency (now TEED) at monthly intervals and independent evaluations took place mid-way through and at the end. A final evaluation report will be submitted to TEED and full dissemination of the project results will follow in due course. Dissemination of the progress of the project has been an integral part of its remit. Project staff have given presentations for REPLAN/NIACE, FEU, Regional Advisory Councils, HMI, TECs, ET providers and numerous colleges.

The individual candidate's progress must be continually monitored by the APL adviser and the administration staff. Feedback is both formal and informal and is constantly updated.

Conclusion

We hope that through our work we have established that APL can be a great opportunity for many adults, including the unwaged. It is particularly effective at raising confidence and self-esteem which may have been eroded by unemployment. Not only can a qualification be of use when seeking employment, but a portfolio which contains evidence of competence can also be of great value. Some clients find the experience sufficiently motivating to seek a course after accreditation.

It must be noted that experience need not have been gained through paid employment and, therefore, those who are out of work will benefit further as generic competences are defined.

Centres wishing to provide an APL service will need to pay close attention to quality assurance. Cost effectiveness can be partly addressed through appropriate staffing and partly by achieving an operating scale which maximizes resource utilization. Good support materials promote independence in many clients. In terms of marketing, it might be expected that the initial response to a completely new idea would be poor. However when an awareness of the poten-

PART III: RESPONDING TO THE CHALLENGE

tial benefits was fostered, people were not slow to respond. It is interesting that many of our clients have stated that they would not normally have considered joining a college course.

Obviously there are still outstanding issues. Although the process need not be expensive in comparison with courses, it is necessary to ask who will bear the cost in the future, and for those who cannot afford to pay. Our marketing efforts have been successful but we must find ways of approaching many who are still not involved. Some people who can claim genuine competence may need additional support where their basic skills will not allow them to successfully gather all of their own evidence. Support mechanisms and group work can be made available, but again there are costs. Some clients may need support to maintain motivation through personal crises.

APL has already won many friends inside educational support and guidance services. However, there are still many others to be convinced and, unless we succeed in changing their attitudes, many adults will miss out on their entitlement to APL.

8 Managing Employment Training

CONTRIBUTORS: PETER BODEN AND SUE ROSE, NORTH DERBYSHIRE CHAMBER OF COMMERCE

North Derbyshire Chamber of Commerce's training practice has evolved and changed in order to meet the training needs of its customers and lead bodies. Its ET, YT and customized training provision has evolved from being a tutor-led scheme to an open-learning style with all of its material produced in modular form. At first it was not easy for tutors to release control of learning. However, once their initial fears were dispelled, the transition from tutor-led to learner-led materials took place and the open-learning (OL) centre developed. It is now an essential, integral part of provision. For the past two years, a fully roll-on, roll-off system of learning has been in operation, thus enabling all age groups and most learning abilities to be integrated into classes. Experience has shown that learners respond positively when given opportunities to be responsible for their learning and the speed at which it progresses.

Description of the process

ET trainees start on 15, 20 or 30 training hours, depending on their circumstances, and most are allocated around six to eight hours in the OL centre. The OL co-ordinator works out with each trainee how best to use the allocated time. The centre provides:

- additional material to support work being done in other classes;
- facilities for work on an assignment set in other classes;
- resources for training in areas and subjects not covered elsewhere, eg, retail, stock control, electrical, etc; and
- resources for building on 'academic' qualifications such as improving GCSE grades.

PART III: RESPONDING TO THE CHALLENGE

> Trainees will sometimes reveal that they know their school results don't reflect their true ability. The OL centre fits very well with working at improving GCSE grades, or even doing a new subject. No-one in this building claims to know everything about all the GCSE subjects, but we can provide all the texts, the syllabuses, past exam papers, examples and we can be available as a sounding board. We're talking about people who've realized they can do it ... we supply the resources, they provide the impetus.

Occupational areas

Clerical is the main occupational area covered. This is determined by what's offered by Thornfield Training Centre at large: the OL centre offers back-up, extra practice and extra topics. OL is very well suited to clerical skills: paper, computers and books are the stuff of offices. But many other areas are offered. There's the distributive industries open learning (DIOL) retail package for training in areas such as stock control, distribution, health and safety. There's also a package for electrical trades, but the centre's experience is that this training must be tutor-led and the package can provide a taster only.

Resourcing

> When I came down here I could see that the Chamber was putting its money where its mouth was.

The open-learning centre is truly inviting. Purpose-designed, it is bright and warm, attractively decorated and carpeted and well-equipped. The room is dotted with big healthy plants. The furniture is attractive and varied: there are cubicle desks for individual work, bigger desks for pairs or groups, sofas, comfortable chairs and a coffee table in one corner. Providing a decent environment with decent equipment has proved to be cost-effective.

> We're very heartened. There are no drawings in the cubicles and, after three years, I think there is only one stain on the carpet, and that was the cleaner. There are lots of personal stereos. We've lost one in three years. They're not booked in or out. People help themselves and put them back.

There's a choice of four different types of computer. In the classrooms upstairs the machines are all the same type, so it's in the OL centre that trainees gain familiarity with the small differences between machines and, more important, confidence that once they can use one they can use others.

Delivering training

When trainees start, the co-ordinator sits down with individuals and talks about what they want to do. That then has to be fitted in with existing resources and a decision about additional resources has to be made. A plan is arrived at.

> I make suggestions. Obviously they defer to my experience and normally agree, but we always emphasize that, if they find it isn't right, they must say so. Because staff don't pose as the experts and because people feel comfortable in the centre, they generally do. In this way, the flexibility of open-learning is being used to help boost confidence. Trainees can be assured that if they want to change there's absolutely no problem. They're not letting anyone down.

Confidence boosting

Trainees are given back-up copies of all software disks; originals are kept safe. This isn't just a matter of looking after precious resources, it's one of the many critical 'little touches' which help confidence building.

> If people are new to computers and I say, 'For goodness sake don't damage that, it's the only copy', what's that going to do to their confidence? I go to great lengths to assure them that it's only a copy.

Learning materials

With learning materials, it was largely a case of discovering what worked through experience. This meant listening and responding to the judgement of users.

> It was new to me as well as to them. We were all learning. Gradually, through their honesty in telling me what they thought, I was able to assess types of materials.

Certain sorts of materials were found to be generally useful.

> People like variety, so I look for material that uses different media; some text, some video, some audio. That helps to keep the learning fresh.

An important factor for resourcing is how far material can be reused. Distributive Industries Open Learning (DIOL) material is a good example:

> You buy the package once, then you can buy separate trainee work-books and assessment modules.

PART III: RESPONDING TO THE CHALLENGE

Cost effectiveness

One of the keys to building up a good resource bank is finding good basic packages. For example, Computer Literacy for Managers is a familiarization course covering the basic functions of computers and cost the centre around £450. The average time spent on it is 25 to 30 hours, and it is used again and again. Well-written packages like this are used by so many trainees that they bring training costs down tremendously, allowing the centre to increase the range of materials offered.

> It's swings and roundabouts. You need good basic packages that can be used again and again. That offsets the costs of the individual packages. We have to offer a full, open service.

Organization of material

Generally, material stays in the centre. Trainees put the material they're using in their individual pigeon-hole. If they ask, trainees are allowed to take textbooks home, say, for the Christmas holidays. No-one wants to dampen that sort of enthusiasm.

An important aspect of this provision is the way it can be tailored to trainees' individual needs and the demands of the job.

> Tutors have a lot to learn from the clients' experiences of work placements. I encourage the learners, when they go out on placements, to look around and see what the company's doing, and how, what it's using and how they might fit in. When they come back and tell us, we can restructure part of their training to take into account what they've found out. If it means buying additional support material, then we'll do it.
>
> Open-learning gives that flexibility. You can take positive strides in the trainees' direction. I'm not telling them what is good for them, they're telling me what they want based on what they've discovered in their world of work.

The centre sees many benefits to this method among which are:

- it's an immediate way of tailoring training to local employers' current needs – not statistics, real perceived needs; and
- trainees are reassured of their worth when tutors' attitudes are 'let's see what materials we can find for you' – it needn't be too expensive, given a balance between packages which will be used a lot and those that will be used only occasionally.

Role of staff

The centre staff act as facilitators, advisers, guides, friends, rather than experts, tutors or assessors.

We're not experts in all these subjects, even if we wanted to be. But we can listen, support and advise. I do get to see external tutors' remarks on assignments, but that's entirely up to the student. Normally I get to see only the good ones ... not so often the bad ones. That's fine. I want to see the good ones. But I am shown the not so good ones sometimes, and that's very useful for monitoring the quality of the materials and tutorial support being given.

The staff's desks are very much 'part of the furniture', not set apart, and the two of them are in different parts of the room so that trainees can easily approach the person they choose to talk to. The co-ordinator's desk is just inside the door so that he/she can greet people and say goodbye. 'It's very important that they know we're always bothered.'

Staff spend most of their time walking round and sitting down to talk with individuals and groups. Normally, around 25 to 30 trainees pass through in a day. Records are kept as concise and simple as possible, though the co-ordinator does find it useful to have records of where trainees are during the week.

If the material we've ordered for an individual arrives, I can look up, see where they are and go straight there at the end of that class to say the material's there. It's another way of saying you're thinking about the trainee ... that it matters.'

Motivation

Motivation is a major issue with open learning. Trainees have to see that staff are interested in them. The onus is on the staff to know exactly what each individual is doing and offer continual encouragement. Staff have to pick up quickly any problems with level of material and offer alternatives.

Work placements

Thornfield was 'a victim of its own success'. So successful was it in providing opportunities for learning, building trainees' confidence and encouraging their aspirations that trainees were reluctant to go out to work placements and the centre was getting 'clogged up'. The centre now insists that trainees take up placements offered after three months or they must leave the training. Two things help.

- Ex-trainees come back to tell others about their successes – eg, the 28-year-old who's never been in work and who has just started doing three days a week in catering. There is no formal arrangement, but ex-trainees call in often enough to have some impact.
- The roll-on, roll-off system necessitates mixed classes of ET, YT, private trainees and employees. Thornfield has found this to be

a totally positive thing with no drawbacks: the older people have a steadying influence on YT trainees, YT trainees raise the spirits of ET trainees, employed trainees bring a breath of hope and insight into the world of work.

Assessment and accreditation

In general, accreditation is found to be motivating:

> The DIOL assessment module is a project which involves going beyond the OL materials. When it's completed, it's sent off for independent marking and there's a certificate at the end of it. That's important. It rewards all the effort.

About 50 per cent of the packages used are linked to certification. The other 50 per cent can be used as springboards to go on to certification.

The relationship between the trainee, the OL co-ordinator, the external college (eg, Pitman) and the college tutor is important:

> A package may have a dozen assignments. Trainees use their home address when they send assignments off. We act as facilitator. The marked assignment with the tutor's comments comes back to them, not to us. That's good. If the relationship between trainee, the college and the tutor is working, that's fine. If they need help, they know they can come to us. Otherwise, I've no wish to pry.

The centre finds Pitman's packages combine the main features of good OL material. Users like it, there is a continuous assessment with good tutorial support and there is accreditation. Marking of assignments is constructive. They start with encouraging remarks. Then, in the analysis, they are frank and precise about shortcomings because they have to be. But at the end there's always an encouraging note.

This case study describes just one specific area of provision. It is not intended to be a full report on all adult training provided by this organization.

9 Providing Training Access Points

CONTRIBUTOR: BOB ADDEY, EAST BIRMINGHAM COLLEGE OF FE

East Birmingham College has embarked on an ambitious programme of setting up education and training access points in the communities of East Birmingham. Its policy is to take 20 per cent of the curriculum into the community, 'taking the college to the people rather than waiting for them to come to the college'. The reason for this is that the college recognizes that the location of its main site is not only geographically, but also culturally, marginal to many of the communities in East Birmingham. This kind of doorstep delivery has proved particularly successful in attracting women into education and training, especially where it is supported by childcare provision. Between 1990 and 1991 ten such centres were developed and existing college resources transferred to resource them. The success of the centres has attracted significant external funding from the city's Economic Development Department, the Education Department and the East Birmingham Task Force. Successful contracts for funded training programmes have also been obtained. To meet the needs of adults in these centres, efforts have been made to ensure that:

- staffing of centres reflects the needs of clients in terms of gender, ethnic origin and multilingual capability;
- language support is integrated with vocational training;
- partnerships are established with other local providers, host organizations, community organizations, and local employers;
- publicity, guidance and counselling are available in community languages;
- the environment is attractive and welcoming to adult learners as well as being culturally acceptable;

PART III: RESPONDING TO THE CHALLENGE

- student services are of a high quality and culturally sensitive;
- core skills (literacy, numeracy, job search, IT) are integrated into all programmes as an entitlement; and
- staff development programmes support community-based delivery.

Programmes in the centres are marketed through: schools; local job centres; a bilingual newspaper, *Community News*; a providers network made up of people from AE/FE/private training agencies/voluntary sectors/careers services; formal and informal representation in community organizations and centre steering groups; and word of mouth.

Because of this approach, 'outreach' has become an increasingly dated concept for the college since it implies the kind of distance from the community that no longer reflects the situation.

The college is about to register for BS5750 and has been committed to a strategy of total quality development for the past two years. This has provided a strong institutional context for the development and management of their own new centres as well as providing a framework for ensuring quality of delivery of all parts of its provision in these centres.

What follows is a detailed description of the provision, how it was set up and what its outcomes are:

Small Heath, Saltley and Washwood Heath form a band of deprivation which is ranked in the worst 2.5 per cent in the country. 90 per cent of electoral districts in the UK are less deprived than the inner city parts of East Birmingham. (*Department of Environment Z score – based on social class, ethnicity, unemployment, overcrowding, lack of basic amenities, single parents and pensioners living alone*)

The college is situated on the eastern outskirts of Birmingham but serves a number of inner-city areas (including Small Heath, Saltley and Washwood Heath) with a large proportion of inhabitants from the Indian sub-continent, primarily from Pakistan but with an increasing number from Bangladesh. This population has grown considerably in recent years and the rise in birth rate has led the local authority to open a number of infant and junior schools in the area in marked contrast to the falling rolls in other areas of the city.

East Birmingham has a higher proportion of pre-school children than the national average – 7.9 per cent compared with 6.5 per cent. (*WMRHA 1988 Final Home Population Estimates*)

It was obvious for long-term planning purposes that these pupils would become college students in the future and it is now college policy to take education and training into the community to service the needs of all age and ethnic groups within the East Birmingham area by setting up centres to deliver new courses and resite existing provision to make it more attractive to the target groups.

The seeds of the project were sown in 1985 when an adult training unit was established in the inner-city. It was followed, in 1987, by a joint LEA/college initiative to set up a community education centre within a newly opened junior and infant school, the eventual success of which led to the establishment of similar centres in other schools designed to cater for the needs of parents and the immediate local community.

> 74 per cent of the Asian mothers had received no formal education and 71 per cent were illiterate in their own language. 83 per cent of the Asian mothers were not fluent in English and 90 per cent stated that they required an interpreter to communicate fully with English speakers. (*East Birmingham Health Authority Antenatal Care Research Project: First Year Report*. J. Dance, C. Hughes, 1987)

In addition the college also developed a run-down YTS (Youth Training Scheme) unit to accommodate a wider clientele and negotiated with an Islamic resource centre to share a site for education and training purposes. By the end of the summer of 1991, this had led to the development of the following centres.

Henley Street, Sparkbrook – a motor vehicle engineering training centre in a free-standing industrial unit, ESOL (English for speakers of other languages) and ABE (adult basic education) support provided.

Parkfield Centre, Saltley – an established centre, on a site shared with a junior and infant school, offering a wide range of clerical, information technology and caring courses, including nursery nurses and provision for bilingual school classroom assistants, ESOL and general education. It also acts as an educational and training advice centre.

Bordesley Green Centre, Bordesley Green – a free-standing centre, designed for full disabled access, offering dressmaking, art and design, garment manufacture, carpentry and joinery, painting and decorating, multiskills, ABE and ESOL. It also acts as an educational and training advice centre.

Golden Hillock, Small Heath – rented from the Islamic Resource Centre and providing dressmaking, multicultural design, garment manufacture, ESOL and ABE support, and courses for home-workers.

PART III: RESPONDING TO THE CHALLENGE

Small Heath Centre – part of a leisure centre shared with the City Recreation and Community Service, it affords some youth and adult provision, with the college running office skills, new technology and general education courses.

Leigh Road, Washwood Heath – part of a junior and infant school, it offers courses in caring (including childminders), office skills, new technology, ABE, ESOL, and an educational and training advice centre is planned.

Park View, Washwood Heath – part of a secondary school, it runs courses in business and finance, office skills and engineering, it is the base for redundancy courses and shares one room with a local job club.

Shaw Hill, Alum Rock – part of a junior and infant school, providing community education courses, ESOL, ABE, access to public services courses and introductory courses in office skills and new technology.

Hodge Hill – part of Hodge Hill Girls' School, it offers courses for women into technology and general education courses.

The Firs, Castle Bromwich – a centre, linked to a junior and infant school, offering general educational opportunities.

Byng Kenrick, Kitts Green – part of a secondary school, it provides courses in caring and general education.

St Peter's College Assembly Hall – the old chapel of an old training college, this centre specializes in garment manufacture in a simulated commercial environment.

East Birmingham Business and Management Centre – again housed within St Peter's College, it is designed for the delivery of business and management courses for the community and local industry.

The Jet Shop, Alum Rock Road – a joint initiative with a number of other training providers in the area to set up an inner-city recruitment agency and advice centre, with the college organizing the use of a training room.

Main Site, Garretts Green – is the main administrative centre, offering a full range of education and training courses.

Efforts are also made to ensure that all student support services are offered on all sites and that the staffing reflects the needs of the clients in terms of ethnic origin and gender, with many of the staff being multilingual and trained in counselling skills. In addition the staff are continually looking to provide a better service by monitoring and evaluating existing courses through established quality procedures and by responding quickly to demand.

> Unemployment is very unevenly distributed across the city with inner-city parliamentary constituencies experiencing both the highest absolute numbers and the highest rates of unemployment. These constituencies are Small Heath with an unemployment rate of 19.8 per cent, Ladywood with 16.6 per cent and Sparkbrook with 16.3 per cent compared with an average Birmingham MDC rate of 10.9 per cent. (*Labour Market Information Bulletin, April 1991, published by Birmingham City Council Economic Development Department*)

Some of the centres have been targeted as local access centres to be equipped with public access databases to provide instant information on demand regarding educational and training opportunities in the area.

The resources and courses within the centres have been funded in a variety of ways but it has been apparent that the proactive approach of the college has attracted funding in the first instance to enhance the physical resources of the centres to a high standard, and only later to develop new provision. Some of the courses are traditional FE courses that would have been delivered on the Garretts Green site but have now moved to the inner-city centres, others are negotiated through YT and ET managing agencies, particularly the Birmingham City Council Economic Development Department. Some provision has also been supported through the European Social Fund, Birmingham Heartlands initiatives and the East Birmingham Task Force. Many of the courses are vocational, with built-in language support to meet the needs of the adult Asian community and there has also been considerable success with employer-led recruitment compacts. The college also runs courses under the Employment Services-funded 'Options' provision in English and community languages.

The new centres have also attracted the attention of local schools, the Social Services and institutions like local hospitals, all of which have asked the college to arrange training packages for their students/trainees. One fruitful outcome of these arrangements involves a local adult training centre providing catering services in a number of centres, with the trainees attending day-release provision in the college, leading to accreditation through Caterbase.

There has also been a policy in east Birmingham to encourage networking and partnerships and the college has responded positively to this. In some cases this has meant joint marketing ventures such as an audio cassette (one side English – one side Urdu), produced professionally with the assistance of the local BBC radio station, that is distributed free within the local community and promotes local education and training possibilities with all local providers, using one of the college centres as the central information point. Another working example is a Birmingham TEC Ltd-funded project called 'Prospects' which involves the college being the access point for a programme involving the participation of four training managers to actively encourage the local recruitment of people into education and training.

The outcomes of the college's community education and training venture have been that:

- the college's profile within the local community has been raised;
- the college is seen as a positive deliverer of education and training, one that cares about the needs of its customers and clients;
- the college is trusted by the community and also by other education and training providers in the area, creating an ideal climate to nurture further partnerships and networking relationships; and
- more adults and, increasingly, younger people from the local community are being encouraged to take up educational and training opportunities and progress to higher qualifications where appropriate.

The success of the venture can be measured by the growing demand for staff and space. At certain times, this demand cannot be met because the centres are full. However, this may encourage further partnerships in the future.

It is now apparent that a venture that was begun in response to the needs of the long-term unemployed and unwaged within the inner-city, many of whom had difficulties with the English language, has now been developed into a proactive plan to ensure the college integrates with the communities it serves well into the 21st century.

10 Supported Training in Employment for those with Learning Difficulties

CONTRIBUTOR: JOYCE THOMAS, TAMESIDE COLLEGE OF TECHNOLOGY

This was a pilot project which operated in the Manchester area from March 1990 to May 1991 and was funded by Moorfoot, TEED, Sheffield.

An effective approach to training and employment for people with learning difficulties and disabilities has been established through the department of Management and Training Services in Tameside College of Technology. It offers a blueprint for securing job opportunities by promoting the belief that all people, irrespective of their degree of disability, have the potential to develop, learn skills, and to get and keep a job if they are offered appropriate training and support in the early stages of employment.

Background

When looking at what has been achieved by national and local training initiatives over the past few years, it could be seen that they have been less than effective in meeting the needs of those with learning and/or behavioural difficulties. This was shown most clearly by the lack of measurable outcomes in terms of trainees achieving employment (5 per cent) and vocational qualification (2 per cent).

It would seem that staff were employed for occupational skills rather than their ability to work with people who have learning and behavioural difficulties, thus creating a stressful environment for both staff and trainees, often made worse by a lack of resources and antiquated buildings and equipment.

PART III: RESPONDING TO THE CHALLENGE

Bearing in mind that most staff have been supervisors/instructors in sheltered workshops for many years, it was not surprising that there was a lack of an employment-led culture, which resulted in them being as institutionalized as the trainees.

The general consensus of opinion among managers and their staff indicated that the existing model did not work and, when compared to any quality ratios, the provision was found lacking and had the following negative characteristics:

- *Transferable skills* – many people with disabilities or learning difficulties found it extremely difficult to transfer a skill from one setting to another or, indeed, use the skills and knowledge in a different context.
- *Dependence rather than independence* – many disabled people had spent a considerable part of their life in situations where they were very dependent on other people. Sheltered workshop environments in many instances encouraged a continuation of this dependence rather than the development of independence.
- *Attention-seeking behaviour* – negative, attention-seeking behaviour was aggravated or copied in situations where peers were adopting similar behaviour patterns.
- *'Getting ready' theory* – some workshop staff imposed their beliefs that people could not go into employment situations until they were 'ready' when, in practice, employers are the best judges of who can or cannot work for them.
- *Assessment* – it was often assumed that it was necessary to process individuals through complex and time-consuming assessment processes which, in many instances, only informed people of what they *could not* do, not what they may be able to do. Most people were aware of what they couldn't do; what they needed was a clear indication of their capabilities.
- Many workshop managers expressed the view that prolonged stays in sheltered/training workshops often correlated with the difficulties experienced in keeping a job, ie, the longer a person was in a workshop, the greater the likelihood of the job placement breaking down after a short period of time.

In other words, training workshops had become the dumping ground for 16- to 21-year-olds who were not readily employable and whose benefit entitlement had changed. The low morale and high turnover of staff was related to the negative aspects of unsuccessful work placements, behavioural difficulties and the 'sheltered environment' approach. Also, it was clear that performance was not going to match up to impending payment by outcome rules. New structures and operations in the training workshops have now been developed which have positive expectations and aim to break the 'dependency' culture which those attending so readily develop and which, in turn, works against lasting, successful employment.

The new model was established based on a restructuring of the existing workshops and providing appropriate staff development to create four new key roles, different from those traditionally held by staff and tutors. These were 'job developer', 'job coach', 'job profiler' and 'behavioural mentor'. The issues which have the greatest impact on people's employment were also identified. These included:

- pre-employment preparation of prospective employees for a specific job;
- accurate job profiling of clients to ensure that the right matches between people and jobs were made;
- consulting the client at all stages of the decision-making process; and
- attention-seeking behaviour decreasing as skills increase.

The overall aim of the scheme was to bring about changes in attitudes towards the training and employment of people with learning difficulties and disabilities so that:

- a basis for developing an alternative approach to the present model was established;
- new thinking on the validity of assessment and profiling could be developed; and
- training and support to scheme staff, before, during and throughout the transition period, could be provided.

Participants

The following organizations and schemes were recruited to participate in the scheme:

Alan Graham Trust (formerly ICAN): (Staff involved = 2) An employment training programme based in Stockport.
The Rathbone Society: (Staff involved = 4) A youth training workshop scheme based in Stockport.
Tameside Training: (Staff involved = 5) A youth training workshop scheme based in Tameside.

Oldham Training Workshop: (Staff involved = 3) An integrated ET/YT programme based in Oldham.

Specialized Technical Services: (Staff involved = 5) An integrated ET/YT programme based in Stockport.

All of the schemes recruited make provision for young people or adults with learning difficulties and/or disabilities ascribed to a variety of reasons.

To prepare the participants for the new scheme, guidance programmes were introduced. A three-day training course for job coaches was carried out to develop job- and task-analysis techniques and systematic instructor skills appropriate for clients with learning difficulties and disabilities. A two-day training course on marketing skills was run for those responsible for obtaining employment opportunities. Also, job profiling and working with difficult behaviour were carried out on an individual basis.

Financial assistance was made available on a four-monthly basis to all organizations. Typically, the funding was used as follows: three organizations employed job coaches; two organizations updated systems, produced marketing material, etc.

Also, each of the participating organizations received consultancy back-up to assist in setting up a supported training in employment (STIE) model. This consisted of in-house training/coaching, organizing the production of marketing material/mailshots and arranging meetings to update the 'state of play', thus allowing the organizations to support each other by exchanging ideas and sharing information.

The STIE model in operation

What follows is a description of how the established model works in practice.

- Trainee/client enters scheme/organization.
- Trainee/client is interviewed by job profiler to complete job profiling form. This information is crucial to the success of the job matching process.
- Job profile is discussed with trainee/client, job developer, job coach and job profiler.
- Depending on the outcomes of the above meeting, the job developer identifies job openings, job tasters, or work experience to carry out situational assessment.

- Job developer undergoes an interview, on behalf of the trainee/client, with the identified organization, and makes arrangements for him/her to start work, or for him/her and the trainee/client to experience job taster (half or full day).
- Job coach carries out job analysis and identifies possible co-workers.
- Job coach undergoes training to do the job and makes him/herself known to everyone, ie, why he/she is there.
- Job coach carries out task analysis.
- Job coach makes arrangements on suitable starting date for trainee/client, then informs all concerned.
- Job coach instructs the trainee/client on-the-job and, if necessary, gives appropriate training away from the job.
- Job coach carries out data collection to inform trainee/client on progress.
- Job coach begins the fading procedure, the timing of which is crucial to trainee/client's success.
- Job coach is now ready to take on another trainee/client.

Key support roles for STIE

Job coaching:
- job and task analysis;
- analysis of job environment;
- briefing and training work-place staff;
- instructing trainees on-the-job;
- supporting trainees in the working environment; and
- evaluating and recording trainees' progress.

Job developer:
- analysing local labour market opportunities;
- producing and implementing a marketing strategy;
- matching trainees with employment opportunities; and
- working with other key support workers.

Evaluation and job profiling:
- interviewing trainees to match job profile;
- work-place assessment and recording;
- implementation and re-negotiation of personal action plans;
- working with other key support workers; and
- working with outside agents, careers agencies, social services, etc.

Behaviour mentor:
- assisting with the development, managing and implementing of the organization's policy for handling behavioural difficulties;
- working in a team to ensure effective implementation of policy;
- considering and supporting colleagues and trainees to effect an agreed approach; and
- devising and monitoring systems for the effective implementation of policy statements.

Recommendations for future users of STIE

- All staff in any organization need to be informed of what the STIE model is, how it differs from present procedures and what (if any) their roles will be. They need to be constantly updated, regardless of their involvement; otherwise, people feel threatened, and become unco-operative. Without this co-operation and team-work, the staff implementing STIE in its initial stages cannot operate efficiently.

- Sheltered workshop supervisors cannot operate as job coaches on a part-time basis – there is too great a temptation to put the workshop first, to the detriment of the trainees. Once a supervisor has been identified and trained as a job coach, he/she must not be responsible for any trainees operating in the workshop.

- Emphasis must be placed on the importance of completing a job profile to ensure accurate job matching; otherwise it will lead to the loss of possible employment opportunities.

- Care should be taken in assuming that the transition from placement officer to job developer/workshop supervisor to job coach would be automatic.

- Some job coaches will be reluctant to get involved in negotiating pay or producing an individual package for their trainees. They may feel ill-equipped because they lack information regarding funding or, mainly, lack the skills and confidence to venture into unknown territory.

- Staff need a sense of security to enable them to progress through the change of process.

- People obtain and lose jobs throughout their working lives, and support mechanisms should acknowledge this. The loss of a job should not necessarily be seen as a failure.

- A 'greenfield approach' is the most desirable in terms of staffing and operating STIE.

- Staff need to be chosen very carefully before embarking on the STIE model.

To sum up, the STIE model gets people out into the work-place as soon as possible and gives them a clearer view of possibilities, assuring them that problems can be solved. The jobs were seen to be 'real jobs', ones that they could carry out satisfactorily. They did not have to cope with the kind of hidden agendas that can often be the cause of employment breakdown for people with learning difficulties. The work was initially a six-month pilot but has been extended and has support from Stockport Training and Enterprise Council. Although it is marginally more expensive to run than the old system, and requires investment in staff training, it gets results.

11 Delivering Adult Compacts

CONTRIBUTOR: DIANE HARRIS, LOUGHBOROUGH COLLEGE OF FE

Introduction

Adult compacts have come under a great deal of criticism from people who believe that the needs of the students must always come first. This is because, if simply taken at its face value, a compact seems to be completely employer led and, therefore, totally in their interests. Organizations which enter into compacts with local employers are sometimes even accused of exploiting the adults who become involved.

At Loughborough College, however, those tutors who are involved in the compact scheme are totally committed to meeting the needs of the individual and this priority remains paramount in their minds when they are working with their students.

Also, it must be remembered that people who enter into the compact scheme do so because they are specifically looking for work. What then happens is that the needs of each particular student are matched with the needs of the employer, and a compact is formed. The whole exercise then becomes a partnership which simply wouldn't work if it was completely one sided.

For a compact to be effective, students have to be given a great deal of support and guidance, their previous experience must be acknowledged and credited, flexible methods of delivery must be adopted and, above all, the student must be motivated to learn.

All of this is recognized as good practice in terms of working with adults and therefore has the effect of increasing the student's confidence so that a lifelong interest in learning is developed. Tutors at Loughborough find that their students not only gain the employment which was their original aim, but that they often want to continue to learn and study afterwards.

Therefore, the following description of the compact scheme at Loughborough College must be considered in the light of the commitment of the staff to put the needs of their students first.

Loughborough adult compact

During 1990 a project which looked at the feasibility of adult compacts was funded by the Further Education Unit and undertaken by Loughborough College. The study proved valuable in examining the attitudes both of employers and individuals seeking to return to work. The need to match skills shortages with local employers' recruitment difficulties and the demographic changes taking place within the school-leaving community were also highlighted within the research.

Statistics received from the Department of Employment identified a skill shortage in secretarial, clerical and administrative work within the community. It was also established that women returners formed a high proportion of those who were seeking to return to work in an office environment.

Loughborough College has been providing advice and guidance for unwaged and unemployed people within the community for more than five years. A close network has developed between local agencies involved in the provision of education and training for adults, and on a national basis. The concept of an adult compact sat comfortably within the Adult Training and Learning Support Unit remit which aims to provide a service to all adults who wish to re-enter a formal learning situation.

The Loughborough adult compact has developed into a tripartite relationship, built on a spirit of goodwill, between employers, Loughborough College and unwaged people seeking to return to work.

The aim of the adult compact programme is to train or re-train adults for clerical, secretarial and administrative work in close collaboration with employers within the community.

During the summer of 1990, as part of the feasibility study, an initial pilot programme took place. The participants were recruited from people who wanted to return to work and had already approached the college for advice and guidance on course availability. This group comprised five women and three men, with an age range of 22 to 62 and a mixed occupational history; all the men and three of the women were long-term unemployed. Of the eight people who joined all completed the full 12-week training programme together with a period of work experience. Significantly, all of the women were able to find work either on a paid or voluntary basis. The three men were unable to get work by the end of the programme but were keen to continue to improve their computing skills.

The success of the pilot compact in building the confidence and self-esteem of those people who had been long-term unemployed was difficult to quantify. However, feedback from employers and participants alike indicated that, in terms of the skills required for re-entry to employment, the aims of the compact had been achieved.

Funds were provided by the EDEC committee at County Hall for three adult compacts to run from September 1990 to June 1991.

The target people were 'women returners' who wanted to return to work as soon as possible. An article was published in the local press giving information about the forthcoming Loughborough adult compact programme and, within days, sufficient people were recruited to start on an intensive 12-week programme, commencing in the autumn term, which included two weeks' work experience. The sessions were run between 9.30 am and 3.00 pm each day to make it easier for those with children of school age to participate.

The curriculum consisted of no more than 20 hours in college to meet the criteria required by the Department of Employment for those participants who were receiving benefit.

The timetable comprised:

- three hours' keyboard training, which includes audio typewriting;
- two hours' wordprocessing;
- five hours on information technology – theory as well as practical sessions using wordprocessing, database and spreadsheet software;
- two hours' bookkeeping;
- one and a half hours' office practice in the training office – photocopying, duplicating and collating, which is requested by staff and students throughout the college;
- one and a half hours of training in telephone techniques, which took place at the Royal National Institute for the Blind Vocational College which is sited on the campus;
- two hours of tutorial sessions – confidence building, action planning, job search and interview techniques;
- one hour of presentations by outside speakers from local employers, both from industry and voluntary organizations; and
- two hour options.

The training focused on the use of specialist tutors and equipment. A base room was found for tutorial sessions and private study and to enable the group to meet in informal surroundings on a regular basis.

A great emphasis was placed on creating good group dynamics, thereby enabling individuals to become more confident and share their experiences with their fellow participants in a friendly and supportive atmosphere.

Many of the staff involved, both as tutors and in a supportive capacity, had worked with adults for some time. A programme manager acted as co-ordinator and liaised with staff, trainees and employers alike to ensure that individual needs were met. The collaboration and work experience with employers within the community was the remit of the programme manager. She was responsible for ensuring that the course team worked efficiently and effectively and was accountable to the funding providers. A monitoring and review process was developed to provide the necessary information for evaluation purposes.

Work experience was negotiated with each individual and an employer by the programme manager. An initial dialogue took place to determine the requirements of the employers and those taking part to ensure that these needs were met on both sides. An interview for the work experience took place with the programme manager present. It was emphasised that the individuals must make their own arrangements in terms of the hours and dates when the work experience was to take place. During this time, a visit was made by the programme manager to get feedback from all concerned. Trainees were also asked to write a short report so feedback could be formally collected.

A personal tutor, who also teaches a specific skill area, provided and collated information for participants and college staff. This information included the external accreditation available, assessing and reacting to the individual personal and training needs, responding to job vacancies and providing support and encouragement throughout the programme.

Loughborough College provides 'records of achievement' for all full-time students and this initiative was extended to include Loughborough adult compact participants in September 1991.

Monitoring and evaluation of the programme has taken place regularly. The tutorial sessions have been used to provide feedback each week, and review meetings with the programme manager and personal tutor have occurred weekly. Formal team meetings have been arranged before, during and after the delivery of the programme. Individual tutorials with the programme manager have been arranged twice during the programme as well as on demand.

Review and planning meetings with employers involved have been arranged prior to the commencement of the programme. Informal discussions have taken place during the programme but it is intended that in the future all employers participating will be invited to comment and provide suggestions for improvements.

It is anticipated that a questionnaire will be forwarded to each participant at six-monthly intervals to enable the programme manager to monitor their destinations.

During the period April 1990 to June 1991, three groups took part in the Loughborough adult compact and a total of 47 people participated. All participants completed the intensive training course and did at least ten days' work experience with local employers, either on a part-time or full-time basis.

At the time of writing a significant proportion of participants have gained employment. Action planning has revealed in most cases that family commitments provide constraints on availability and timing for women returners. Participants often need to take temporary part-time work to enhance their chances for more responsible positions in the future.

It has been apparent that the Loughborough adult compact has produced 'graduates' who are keen, motivated and committed. Many of the aims of all those concerned have been achieved and it is envisaged that the concept of adult compacts in other skills shortage areas will be explored and developed in the near future.

12 Guaranteed Accommodation and Training for Employment (GATE)

WRITTEN BY CLAIRE LEVY, FREELANCE CONSULTANT/TRAINER, FROM ORIGINAL MATERIAL PROVIDED BY JULIA CARTER OF THE LONDON ENTERPRISE AGENCY

Aims/objectives

Of the 50,000 to 60,000 homeless people in London, about 2000 live in cardboard boxes and the rest shift back and forth between hostels, squats and friends' floors. Now, 20 major firms have pledged to break the cycle of insecurity, despair and deprivation. The London Enterprise Agency (LEnTA) has taken 25 of the City's young homeless off the streets into a pilot project, GATE (Guaranteed Accommodation and Training for Employment), to provide accommodation, training and work.

The project aims to break the vicious circle of 'no job, no home, no job' by focusing on both employment opportunities and accommodation. Clients are offered a training compact (job-related training with a strong likelihood of employment or even a job guarantee) underpinned by hostel accommodation, intermediate and permanent housing.

The training compact thus seeks to ensure that clients can be recruited into permanent jobs with realistic prospects for future advancement. 'The clients then can become self-sustaining members of the community.' Connections between accommodation and employment are seen as critical for the clients' well-being and for their long-term chances of success.

One of the main aims of the project is to enable people to move more quickly from hostels, via intermediate accommodation, into independent housing.

PART III: RESPONDING TO THE CHALLENGE

Target group

- The young homeless, priority being given to 18- to 25-year-olds. In reality, this extends up to the mid-40s.
- Under-18s are catered for in a special project to take account of their low earning potential and housing benefit regulations.
- The 'not roofless', who have to be in temporary accommodation for the duration of the course.

Funding arrangements

The project is funded by the private sector members of LEnTA for one to two years. Matched funding was received from the Employment Department and City Action Team for training courses.

The partners

LEnTA finds an employer who offers the job guarantee, does a course outline and organizes funding and negotiations with housing projects. IBM has taken the lead on the GATE steering group, which includes Shell UK, John Laing, Grand Metropolitan, Citibank and British Telecom. British Rail was also involved and was looking to improve the recruitment of young people. BR expected course participants to complete the same aptitude tests as any other job applicants.

The training manager was Citicare; the training provider, Kingsway Centre.

Description of curriculum/content/delivery

The key curriculum areas are team building, assertiveness, interviewing skills, literacy, numeracy and timed test practice.

Content
College based:
- personal and social skills – confidence development, team building, assertiveness;
- literacy and numeracy – preparation for tests;
- work preparation;
- preparation for selection – interview and application form practice; and
- attendance and punctuality – compact goals

Work-based element:
- awareness of the organization;

- familiarization with a range of jobs within an organization; and
- application and interview experience.

Delivery – four-week period
Timetable

Week one: Induction and registration; preparation of aptitude tests/application forms.

Week two: Aptitude tests/application forms continued; preparation for interviews.

Week three: Interview preparation continued; job selection.

Week four: Interview preparation continued; interviews.

Outcomes of GATE pilot review

Employment:
Of the 24 trainees recruited onto the course, 23 achieved their goals and were offered employment at the end.

Project time/developments

The GATE pilot was run as two courses, ie, March/April 1991 and May/June 1991. The main employer partner in these was British Rail.

Changes were made to the second pilot as a result of feedback from the first. These were:
- closer tie-in with recruitment *process* to avoid build up on the last week of the course;
- course more clearly *focused* and broken down into tasks which were completed on a week-by-week basis;
- course content separated into three areas taught by different specialists; and
- organizational input from British Rail remained but familiarity led to improvements.

Additional changes for subsequent courses were slight amendments in the recruitment process, as suggested by housing partners, and provision of a short pre-interview refresher course for trainees who are not immediately employed (to avoid confidence draining).

Key aspects

The critical factor is the *guarantee*, which gives enormous security to this very vulnerable group. In this case the guarantee is double edged: job and housing.

Also fundamental to success is the *commitment of the training manager* who liaises with the employer on a day-to-day basis and supports and coerces the trainees.

The *involvement* of British Rail in selection, training, visits, etc was another key element.

Assessment/recruitment patterns

Initial assessment – application form:

- evidence of commitment to employment and scheme was demonstrated by completing form;
- evidence of having experience of work and shift work; and
- successful study or training.

Second assessment – interview:

- to demonstrate commitment to project/employment, short written piece to identify trainees in need of literacy support.

Trainees had to have a reference from their hostel which confirmed they could stay for the duration of the course.

The students

John, aged 20, was always early. He did not have enough money to buy a clock so, rather than fail GATE's requirement for punctuality, he set off for college once it was light. After taking the first exam of his life, he started to believe in himself and is now on track to becoming a train driver.

Julia Carter, homeless project manager for LEnTA, says: 'The key is the guarantee. These people had been let down by everyone in their lives so far. We said that if they meet simple criteria, including 100 per cent attendance on a four-week course, they would be guarenteed accommodation and jobs. The work had to offer prospects for promotion and pay £200 a week.' According to Julia Carter, 'There is no point putting homeless people in casual low-paid jobs because it perpetuates the cycle of deprivation. Yet the jobs had to be relatively unskilled, as a long period of training would not be good motivation.

Future developments

Key points
The pilot was successfully completed. British Rail wishes to continue to be involved in the project.

GUARANTEED ACCOMMODATION AND TRAINING FOR EMPLOYMENT

The housing side is the main cause for concern because it is affected by recession and lack of job movement. There are future possibilities for housing developments funded by public and charitable sources in South London. Similar possibilities exist for working with other housing associations on a mixed accommodation, training and small business scheme. (Schemes provide a permanent basis for GATE links and initiative with possibilities of permanent funding.)

The French hostel system ('foyers') which, like GATE, combines accommodation and training is being studied with a view to applying the model to the UK. The working group, led by Grand Metropolitan Community Service and Shelter, is co-ordinating progress.

Within two years, the Agency's associated property division, together with the Peabody Trust, hopes to have a purpose-built building for 200 young people in South London. This will incorporate hostels, a training centre and move-on accommodation. Already, £7.6 million has been pledged towards its cost, including at least £1.6 million from the Government.

Aspirations are high. Although LEnTA cannot fund everything, the final goal is 'jobs and homes for 10,000 of the London homeless who are good, competent and employable'.

13 Hitch-hiker's Guide to Science and Technology: An Access Initiative

CONTRIBUTOR: BRENDA FULTON, SUNDERLAND TUC UNEMPLOYED CENTRE

The initial steps

The Sunderland TUC Unemployed Centre was approached by the Continuing Education Department of Sunderland Polytechnic with a view to discussing a community-based project which would encourage unemployed adults to sample informal education within the sphere of science and technology. A pilot course, run the previous year by Durham University and called 'Women make sense of science', had been successful in attracting women into this non-conventional area and, building from its success, we hoped to provide a similar course in Sunderland.

From the outset it was decided that this must be a truly collaborative affair, linking in informal educational providers as well as the statutory providers of education. A recognition of the fears and doubts of the unemployed about returning to education ruled out Sunderland Polytechnic as the sole provider of such a course. Indeed there was to be a conscious effort to 'outreach' this project to make it 'people friendly'.

We were sympathetic to the view that science and technology was often viewed with awe by the lay person, and to the experiences many had in school of experiments with chemicals and litmus paper which had little or no relevance to real life. The demystification of science and technology as a process was essential in combating feelings of it being 'not for the likes of me'. This course, therefore, would be about making links between science and technology and the real world of work and, hopefully, bring about a realization that career and educational opportunities could be open to all. The wheels were then set in motion to develop a community-based pre-access course in science and technology.

An initial outline proposal of the course with costings was put together by Sunderland Polytechnic, the WEA and the Sunderland TUC Unemployment Centre, setting out aims and objectives, the target group, and the evaluation process. This would be the basis of negotiations with the Wearside Training and Enterprise Council (TEC) to obtain support and funding for the course.

The Wearside TEC agreed to both finance the initiative and to give their support through representation on the management committee. A co-ordinator would be employed to oversee the project and to deal with the day-to-day planning. Administrative support would be supplied via the TUC Unemployment Centre. Additional expertise would be brought in on a consultancy basis, at an agreed daily rate. This would keep the cost of the project to a minimum.

Although very much a raw pilot project, we were conscious of the need to link such a course with real educational and occupational opportunities. Predicted demographic changes and shifts from traditional heavy industry to service sector were taken account of, as well as the skills shortages within existing 'high tech' industries. For the course to be successful, what was needed was a progressive route to real employment, mapping out the various access courses available to higher education. Moreover, each student would be responsible for negotiating his/her own progress and personal development programme.

Consideration was given to following the same pattern as in Durham and making this a course for women only. However, given the high level of male unemployment in Sunderland and the obvious need to attract both males and females into pre-access courses, the target group chosen included any unemployed adults with an interest in this field.

The course was to be run on a pre-enrolment basis, with no entrance qualifications or experience required, although we were aware that students would need basic literacy and numeracy skills so an element of vetting would be necessary to ensure applicants could cope with the course content.

The birth of 'Hitch-hiker's Guide to Science and Technology'

Two key elements were required to attract unemployed people into this field – good publicity and flexibility.

Three separate courses were agreed – two part-time day courses and one part-time evening course. All of them would offer childcare facilities for pre-school age children and timing would be flexible enough to attract those with young school children, ie, 9.30 am start, 3.00 pm finish.

Leaflets, posters and press releases were considered to be the best mode of publicity. Information was sent to the local newspapers as to when and where the courses would take place; also included was a brief outline of the course content along with details of creche facilities and finance arrangements.

A specially designed leaflet was distributed throughout libraries, voluntary organizations, etc, using a contact list supplied by the Sunderland TUC Unemployment Centre and the WEA. As a collaborative affair, the wheels were well in motion.

Enrolment for the courses was overwhelming. For each person who rang to enquire about the course, details were sent out with full information on the course programme, times, meeting points, etc. The initial two part-time courses were fully booked within two weeks of the publicity material going out. The full-time course proved equally as popular and, indeed, was oversubscribed, leaving no choice but to turn people away.

Delivery

Deciding on a course curriculum was to be the joint responsibility of the steering group and the co-ordinator of the project. Anne Johnson, a tutor for the Open University, was chosen as the co-ordinator, and she had some definite ideas on the way forward. Anne presented a full course itinerary, linking personal development, practical skills attainment and field-study visits. It was hoped to introduce students to living science and technology. The field-study visits would include trips to laboratories, companies using modern technological equipment and environmental science projects.

Information technology awareness, development of computer skills, wordprocessing and information handling, would also be important elements of the course. The strategy was to give students a taste of various components making up the whole gamut of science and technology. In that way, several avenues of development would be opened.

The third element of the curriculum would be careers and educational guidance. This would be essential in enabling students to progress beyond the pre-access stage. An effective information network was required, with the back up of the HE and FE colleges, to give up-to-date information on all access courses available within the region.

Running three separate courses in parallel was a nightmare to organize. For example, students from each one had to be given the opportunity to take part in study visits and this proved to be difficult when organizing visits to laboratories because a realistic, workable number for a tour party was between eight and ten and we were often trying to cater for up to 20 people at one time. This meant relying heavily on the goodwill of companies, institutions, etc.

Although it was very much a community-based project, we were aware of the shortcomings in the facilities available and felt strongly that we should give students access to the very best equipment, tutors, materials, etc. Practicalities dictated that we use research laboratories, etc, within Sunderland Polytechnic and other educational institutions for some of the more complex classes that required particular expertise. Wherever possible, meeting points for such visits would be arranged away from the college so that students arrived *en masse*.

Evaluation process

This was to be a two-fold process. Because the original aims of the project were to combat the preconceived ideas of science and technology as being 'not for the likes of me', attendance and enthusiasm would be the first bench-mark of evaluation. Monitoring of this was not a simple task since it depended on honest feedback from participants.

On the last day of each course, all participants would be given an evaluation sheet, which would be broken down to allow each element of the course to be given consideration. Students were also asked to consider whether they were thinking of going on to FE or HE and which, if any, course they would like to attend.

Informal contact with tutors from local colleges was an integral component of the course and allowed students to enquire about specific course content, and the standards required, on a face-to-face basis, giving them the opportunity to make informed choices as to their future path. Evaluation, therefore, could also be based on potential progression through higher and further education.

For a full and comprehensive evaluation of the course, feedback from the tutors directly involved in teaching the course is necessary. To that end, we hope to set up mechanisms to interview tutors and get their personal perceptions of the course and what, if any, amendments are required for future courses.

14 Employment Training Access Course

CONTRIBUTOR: ANN SHOULTS, CRAWLEY COLLEGE

Background

This is a collaborative project between County Training (ET) and Crawley College. For several years the college (like many others) has been involved in developing courses for the adult unwaged, funded by MSC, Training Agency, ET. One of the constraints on the types of programmes that could be offered was that they should be linked to gaining employment (ideally, as quickly as possible). Because funding arrangements were also often tied in with local labour shortages, the challenge here was to produce programmes that improved overall employment prospects as well as emphasizing personal development. The difficulty was that such 'training' programmes were often short term and, while they enabled people to take the first step into employment, they were (usually) unable to offer career advancement. Also, they did not necessarily break the traditional patterns of work for women, ie, the emphasis on secretarial skills, etc.

It is often assumed that the 'adult unemployed' have an identifiable profile. This is not the case since they are not a homogeneous group and have differing needs, experiences and cultural backgrounds. Working in educational guidance, I found that many of the adult unemployed, rather than wanting strictly vocational training, expressed a need for education that would allow them to gain a professional qualification. Our initiative came about through discussions between the college and County Training, both of which recognized the need to extend education and training provision within the locality. Without the vision and support of County Training, the full potential of the education/training link it created would not have flourished.

The project has broken new ground by linking education/training to the long-term needs of the adult unwaged rather than merely filling immediate local skill shortage areas. Both the adult unwaged and the economy will ultimately benefit from a more professionally trained and educated work-force.

The purpose of the access course in general is to provide an alternative route into higher education for mature students who do not possess the traditional entry requirements.

A fundamental aim of the course is to address issues of equal opportunities in education and recognize that certain groups in the local area have not had an equal opportunity to fulfil their educational potential. Many people are capable of getting degrees but were unable to complete their education for various reasons.

Because ET agreed to fund the course in full, the target group was the adult unwaged. Selection was based on an aptitude for communication skills to allow sufficient development in one year to compete on equivalent terms with 'traditional' students, and meeting the usual criteria for ET funding. Students on the course had no fees to pay and received £10 per week in addition to any benefit entitlement, and they were entitled to travelling expenses. Single parents also had childcare fees paid which was a lifeline to several people who could not have contemplated undertaking the course without it.

Providers' names – Crawley College:

Mike Gregory (Head of School of General and Continuing Education, where the course is based)

Ann Shoults (College Access Co-ordinator – also developed the curriculum)

County Training:

In particular, Barry Smith provided unfailing support to both staff and students.

Content

With the needs of mature people in mind, the course runs between 9.30 am and 2.30 pm.

Communication and study skills begin the course and these are designed to develop both the skills and confidence of these students who often have little faith in their own abilities at the beginning of the course. Once they receive feedback on their early work and realize that they might be able to do it after all, their confidence grows and reinforces their intellectual skills.

The course is interdisciplinary with the curriculum comprised of Cultural Studies, English Studies, History, Social Policy and Change and an integrated study. GCSE Maths and English and RSA in IT are offered as part of the course.

The teaching methods are varied: seminars, discussions, micro-lectures, pair and group work, etc. As the group varies widely in education background and experience, the teaching strategies have to be appropriate to mixed ability classes with individual study problems addressed as they arise.

A crucial part of the course is tutorial support and guidance such that each person has time alone with his/her tutor and can be confident that they will receive maximum support in academic work. Clearly a mutual confidence must exist for this to work effectively – thus tutors are chosen very carefully!

The fundamental overall aim of the course is to develop each person's critical and analytical skills, and everything is geared towards this end. The content is important but, at the end of the course, we hope that students have learnt to question any material they may be presented with and have developed an analytical rigour that will stand them in good stead in higher education.

This particular course started in May 1990 and, since each student was entitled to one year's entitlement from ET, it finished in May 1991. By any standard the course was highly successful – a very high proportion of participants gained places in the HE institutions to which they had applied. Just as important to all those involved in the project was the feedback from the evaluation document each student completes at the end of the course – delight at discovering hitherto unsuspected abilities and the opportunities offered by the project were recurring themes.

This was a pilot scheme in this area and, if the outcomes were not successful, then there would not be another course. Even though the successful outcomes exceeded our projections, the continuation of the funding from ET was jeopardized by a large budget cut at Easter. However, we now have funding for another year and the second group has just started. There was an overwhelming response locally and the second course was oversubscribed twice over without advertising – as it is we have 32 people on the present course and feel confident that there will be a large percentage of 'positive outcomes'.

If asked to summarize the key reasons why the project is successful, I would say that it is because it gives unwaged people a very real opportunity to effect a big change in their lives and realize their potential to become professionally qualified people. The importance of the financial support cannot be overemphasized because this alone admits people who, by definition (adult unwaged, often single parents), have few financial resources and are thus often denied access to education and its benefits. Clearly, ongoing support and committed tutors are important but, without the mutual confidence that exists between the college and our local branch of Employment Training, the project could not thrive as it has.

The cynical view of ET funding – that it serves to manipulate the employment figures by taking people off the unemployment register – has often proved a difficulty with those people working with the adult unwaged. Yes, we would have to take advantage of any funding available for education/training for adults but, as practitioners, we are aware that this funding is directed towards the needs of the employers rather than addressing the needs of the adult learner. Because the funding in this project has been directed firmly towards long-term goals, it will eventually enhance both the quality of life for those people who are unwaged and benefit the economy as a whole. It is noticeable, too, that people on this course now have a positive view of ET, seeing it as a benevolent godmother/father rather than a big brother.

15 Community Joblink

CONTRIBUTOR: CHRISTOPHER McCONNELL, FINSBURY PARK COMMUNITY TRUST

Aims and objectives of the project/initiative

Community Joblink (CJ) attracts new employment opportunities to the local community by marketing customized training as a method of closing the perceived skills gap between the long-term unemployed and employers with recruitment needs.

Target groups

Long-term unemployed adults from ethnic minority groups, women/Asian women returning to or entering the labour market.

Funding arrangements

Each scheme is fractionally funded by the contracted employer. Remaining funding is currently derived from the European Regional Development Fund (matched funds from local authorities).

Past schemes have been collaboratively funded by the employer, the Employment Department's ET programme, and Business in the Community. New funding links are being established with local authorities and TECs.

Training provider

Haringey Education Services Training Agency.

Description of the content/curriculum/activity, and how it is delivered

Content/curriculum/activities are dictated by the identified recruitment need of the employer. Though most schemes have been in service industries (financial, retail), the range of employment has varied from sales assistant to management.

In conjunction with an employer and a training provider, Community Joblink develops and manages a competence-based, on-the-job, work-place assessed, customized training programme, normally no longer than 12 weeks in duration. The programme is supplemented by an off-the-job component which addresses skills more effectively acquired when initially introduced by a tutor in a simulated or more controlled environment (ie, classroom, workshop, distanced learning, open learning). Counselling and support are offered throughout.

Outcomes

Every trainee successfully completing the training programme is guaranteed a job. Depending on the duration and vocational area of the course, NVQ accreditation is often included.

Length of time, etc

Community Joblink has been operating since 1987. Initially, when administered collaboratively with ET, difficulties arose as a result of ET's negative impression of some members of our target client groups. Community liaison found that potential applicants were generally sceptical about Government training programmes. The reasons given included: meagre training allowances, exploitative connotations, poor or non-existent employment prospects and protracted training programmes which were viewed as insufficiently 'work specific'.

In response, funding from the ERDF was secured. This allows us to by-pass ET and 'purchase' our courses from providers, pay training, travel and meals allowances, training bonuses for completes and guarantee childcare. With the funding to subcontract training provision, we can identify the most appropriate provider at the best value. In addition, spare ET places for customized training are difficult to find.

PART III: RESPONDING TO THE CHALLENGE

Key aspects of the project

The employer's involvement is crucial to the development of the training programme and the offer of employment. Our clients generally have great respect for, and experience of, training and/or further education as means of skills acquisition. Unfortunately, a concrete transition from the acquisition of skills to the acquisition of a job is non-existent, especially for members of ethnic minority groups and women returning to work. Without the presence of an employment outcome, training for skills and/or qualifications can exacerbate the poor self-esteem of the long-term unemployed. The stigma of unemployment becomes as demoralizing for the trainee as it is traditionally prejudicial for the potential employer.

By identifying and guaranteeing vacancies, CJ legitimizes the employment access outcome of training more assuredly than any other project. By implementing on-the-job competence-based assessment, CJ reassures trainees that success will be based on achievement of employment-related objectives. By attracting employers to an overlooked labour market, CJ broadens private sector perceptions of recruitment, personnel and training.

Job and/or further education and training potential for students/trainees after they complete the project

Trainees successfully completing the training programme are guaranteed a job. When NVQ accreditation has been involved, trainees are also given the opportunity for further post-employment assessment towards NVQs.

Three of our schemes have served to introduce and embed NVQ accreditation in training policy for the sponsor employer.

Assessment and recruitment patterns and procedures

All staff supervisors of trainees are trained in the use of competence-based assessment for customized training. Assessments are done every three to four weeks and are attended by a liaison officer of the training provider, the on-the-job training supervisor and the trainee. Results are fed back to a Delivery Management Group (DMG) made up of representatives from the employer, training provider and Community Joblink. Action points determined at the DMG are based primarily on progress toward successfully completing the training programme.

Applicants are found through liaison with local community groups, the employment service, training providers and local advertising. All interested applicants are booked onto briefing sessions, where up to five applicants will have the opportunity to discuss the scheme with a CJ staff member and fill out an application form. They are encouraged to exhibit transferable skills acquired non traditionally, ie, raising a family, voluntary work, hobbies, casual businesses, etc.

Interviews, held at our offices based in the community, are informal in tone and usually include a representative from the employer, a training provider and CJ staff. Criteria for interviewing are based on both potential and desire of applicants to successfully complete the training programme, rather than the usual employers' standards of employability.

Identified methods for feedback and evaluation

As mentioned earlier, feedback and evaluation is done through the DMG, plus the Employment and Training Subcommittee of the Finsbury Park Community Trust and the Management Committee of the Finsbury Park Community Trust. Trainees meet with CJ staff every week to express concerns during the training period.

Other issues specific to our area of work

It is important to note that although the guaranteed offer of employment is at the core of our work, it is by no means the sum of the project. Our criteria for choosing to work with an employer are based on the community's perceived accessibility of the vacancy. We have rejected employers in the past. To maintain a positive profile with our client group, we must attract employment opportunities from which our clients feel they have been effectively disenfranchised.

We also examine personnel, training and management practices of employers to assess how successful our trainees will be as employees of that organization. Further training and promotional opportunities are vital. We monitor retention of our trainees, and have been asked to advise some employers as to other retention problems they have experienced.

16 Tourism Innovation Project: An Example of International Collaboration

CONTRIBUTOR: LENE BAK, NELLEMANN, DENMARK

Introduction

The Tourism Innovation Project is an international project developed through co-operation between the County of Aarhus in Denmark and Stichting DO-IT at Tilburg in the Netherlands.

The project has two objectives: to transform practice in the domestic tourism and recreational sectors, and introduce innovative vocational training in these sectors. The second objective is met by training Tourism Management Experts and, in the long run, this initiative will also meet the first objective.

The ultimate aim is to introduce a bachelor degree in tourism at the University of Aarhus, which would be based on the experience gained in this project.

A Tourism Management Expert is trained to work in the fields of domestic tourism and recreation. Special emphasis is placed on innovation and improvement of possibilities as well as expected future developments in tourism/recreation.

The Tourism Management Expert training programme must be considered to be a supplement to previous education. It represents an expansion of previously acquired knowledge and skills.

The primary target groups are unemployed university graduates and other persons from higher education. These people are not only well-educated, they are also mature and, in addition, some have gained professional experience.

45 per cent of the running costs of the training programme are financed by funds obtained under the European Social Funds 'Special Action' programme. The remaining costs are covered by the providers of the training programme, or by local foundations (in

part). In Denmark, the remaining 55 per cent is paid by the County of Aarhus which in turn receives state subsidies to cover part of the expenditure.

The Danish provider of the training programme is the Aarhus County Employment and Vocational Training Centre. The Dutch partners are the Dienstencentrum voor toerisme en verkeer ('Research, Training and Consultancy Centre of the Dutch Institute for Tourism and Transport Studies') at Breda and the Bureau of Intermediate Technology at Tilburg.

The Danish and Dutch providers work independently during most of their respective training programmes, and the programmes are not entirely identical. However, both programmes include two exchange visits, and a joint excursion to Brussels, which give rise to some degree of co-operation.

Content and how it is delivered

The Tourism Management Expert training programme runs for 38 weeks. During this period, a wide range of subjects are discussed and treated, and many different training methods are employed:

- lectures given by scientists, business managers, independent professionals and key persons from tourist organizations;
- team assignments;
- case studies;
- trainee period;
- excursions;
- practical skills (eg, using computers); and
- development and execution of a pilot project.

The subjects treated can be divided into four modules:

(1) An introduction module, which introduces the student to the labour market in general, and to the labour market of the tourism trade and industry in particular. This module includes goal-oriented application training
(2) A module incorporating subjects specific to the tourism and recreation sector. This module includes:
    - scientific theories on recreational activities, analysis and statistics;
    - tourism and recreation fundamentals;
    - policies on tourism and recreation;
    - tourism and environmental planning; and
    - international developments in tourism.

PART III: RESPONDING TO THE CHALLENGE

(3) A module on business management. This module includes the following subjects:
- strategic planning;
- marketing;
- quality management;
- business economics/financial management;
- project development;
- fund raising; and
- an introduction to the organization of the EC.

(4) A module which aims to develop and improve valuable techniques and skills such as:
- social and communicative skills;
- personal creativity; public relations and layout techniques; and
- information processing and distribution.

To a very high degree, the training programme is based on the interests of the students and, by way of the training methods, the programme seeks to draw on the personal resources of all of the students.

In fact, personal development is a key issue of the training programme. Traditionally this type of training programme takes the shape of 'pure vocational training', but in this instance the programme is designed to simultaneously strengthen the student's professional skills and her/his personality by means of method and content.

Long-term unemployment causes the unemployed person to become socially isolated and gives rise to both a personal and professional insecurity. Pure vocational training can rebuild and develop the professional skills of the unemployed but the personal and social skills need to be developed as well.

This training programme is based on the attitude that vocational training is good for promoting employment, but it often has an insufficient effect if it is not accompanied by other elements. Leaving out these additional elements would presuppose that the individual unemployed person is in a position to process the content of the programme and to use it actively in a job search process and later on in some future employment situation.

The personality developing module in the training programme starts off in a relatively objective way with a course on communication techniques, job search, etc, and then progressively develops to involve the individual person in a more intimate way.

An unemployed person often sees herself/himself as drawn into a vicious circle of being constantly turned down, and this leads to a loss of self-confidence. If this circle is to be broken, it is necessary to focus on the psychological condition of the individual person. In this connection there are three requirements: openness from the person concerned, a feeling of security in the group and professional tools for processing the experience. The method of achieving this is to make the process relatively long and goal-directed.

The evaluation results of the first training programme and experience gained in connection with previous projects show that the process has been successful.

Potential for students

After completing the training programme, the students are expected either to establish themselves as independent advisers or to join a company specializing in tourism/recreational activities (private or public). In any event they will have become more viable competitors in the labour market.

Project period

The project has been running since August 1990, and the first 20 Tourism Management Experts completed the course in May 1991. A second course began in August 1991.

Partly as a result of the final evaluation of the first course, the second course has been altered somewhat. More time will be spent on studying financial management, there will be fewer case-studies, and the training periods will include a written report from each student.

Outcomes

Now that the first course has been completed it would seem that expectations of its outcome have been too optimistic. Out of the 20 students who entered the training programme, four are now employed by private companies, six have found employment in the public sector, one is planning further education, five have started a workshop and four remain unemployed. None have started their own business so far.

Key aspects of effectiveness

One of the key aspects that make the project effective is the way it exploits the various human resources and the potential of the individual students. In particular, the team assignments give them a chance to bring out their talents, and combine these talents to form a stronger whole. Synergy is thus created.

The international context of the programme is very important, both in relation to the development phase and the resulting training programme. By means of collective research, training and training exchanges, the students are made aware of differences and similarities between Denmark and the Netherlands, and this awareness will also be useful in relation to third countries. A co-operative effort of this kind also helps to create a network to draw on in the future.

Assessment and recruitment procedures

The main recruitment criteria for admission to the training programme are that applicants must have been out of a job for more than a year and still be full-time unemployed, and that they have been in higher education (minimum of three years of theoretical education after the examination for the school-leaving certificate).

20 students are selected on the basis of written applications. No interviews take place. Applicants must give details about themselves, about their educational background, their previous employment and their knowledge of, and possible connection to, the tourism trade and industry. The applicants' personal reasons for wanting to enter the training programme are considered to be very important.

Two fundamental conditions that must also be met are an equal representation of both sexes and as wide a range of ages as possible. Ability to write and speak English is compulsory.

Methods for feedback and evaluation

All the lectures are subjected to criticism by the students and, from time to time, the form and contents of the programme will be discussed with the course manager. This feedback is very important, although only the participants in the next course will be able to profit fully from its results.

However, the final evaluation at the end of the training programme is still more important, as it gives the students a chance to take a comprehensive look at the course as a whole, and see if their feedback has resulted in any adjustments.

Other issues

Co-operation with the tourism trade and industry is very important!

Organizations in the tourism trade and industry should be informed comprehensively about the project. Key people should be consulted for their opinion on the project and the training programme, and the initiators must make sure that their project does not interfere with other similar projects or project plans.

17 APL/APEL and the Empowerment of Adults

THE 'CENTRES INTERINSTITUTIONNELS DE BILANS DE COMPETENCES' IN FRANCE
CONTRIBUTOR: ARMINA BARKATOULAH, INDEPENDENT CONSULTANT, FRANCE

Background

The idea to create a particular place, a 'centre', which adults could refer to for assessment of their prior and experiential learning originated at the beginning of the 1980s, with the socio-economic context of that era. It was then found that workers have to adapt to technological changes and different work organizations and conditions if they want to stay in employment.

As different Government programmes were developed to fight unemployment among adults as well as young people, the job searching services and the training agencies found that the existing systems of recognizing prior learning were inappropriate and inadequate.

This chapter does not give a detailed description of the 'bilan' and all its socio-economic and cultural facets. What will be dealt with is the need for a system which gives credit to adults who have 20 to 25 years of work experience. If these people find themselves suddenly in need of retraining or reskilling, they will need to know what competences they have developed and what their potential is. This fact led to an awareness of the importance of learning from experience and the need to accredit it formally.

It also became obvious that the process of education and training did not end after school or college since a diploma, degree or other form of certification does not demonstrate the possession of sufficient knowledge, skills and competences for a lifetime. The idea of lifelong learning was born.

PART III: RESPONDING TO THE CHALLENGE

Above all, in the present-day economic climate, there is no guarantee of employment for life. Indeed, employment will become a discontinuous process; people will be asked to have a high level of adaptability and greater overall mobility, ie, move from one job to another, from one company to another and from one geographic area to another.

Consequently, people must be able to make the best of the different experiences they have had during their professional, educational and social lives. Faced with rapid changes in the work context, they must be able to:

- take stock of their capabilities;
- be aware of their potential;
- manage their personal resources in the best possible way; and
- anticipate changes and negotiate their way through them.

Underlying the whole idea is the philosophy that any person has the potential for advancement resulting not only from their formal education and training, but also from the various experiences they have accumulated during work activities and social interactions. These experiences build competences that receive no formal recognition because they are not officially accredited. Yet people can mobilize them in a work context or in further training. Formal accreditation of these competences is difficult (in France), but there is a need to establish a form of recognition that would do justice to the individual. This is the rationale behind the policy for the recognition of learning in France. People can no longer rely solely on traditional ratings from formal schooling and must be prepared to master their own professional paths and negotiate the value of their competences. The emphasis is therefore placed on recognition, be it self-awareness or that given through the work context. The creation of the 'Centres Interinstitutionnels de bilans de competences' (CIBCs) is the result of this idea.

What is a 'bilan'?

The word 'bilan' in French summons up the idea of a balance sheet drawn up by an accountant, with columns of assets and liabilities, in the process of taking stock of a company's economic situation. The term in French has also a medical connotation, meaning to give someone a health check-up.

In fact, it is a formative process during which people – with the help of professionals who can identify participants' capabilities, attitudes, strengths and weaknesses – assess their competences so that they are aware of their potential and can use this awareness in their search for future employment and/or training.

It is a global and integrative approach in which the time element is important. The assessment of competences is linked up with the personal and professional plan of the person. The 'bilan' is a living process and one of its main aspects is that the participant must be a volunteer.

The results will sometimes be given in a synthetic form, being written jointly by the client and the specialists or, more often, put in a sort of portfolio called 'porte-feuille de competences'.

The whole process is a combination of intense individual reflection and interviews or workshops with an inter- or cross-disciplinary team of specialists from various fields of education, training, psychology, guidance, etc.

The 'bilan' has a definite psycho-pedagogical slant, the aim being to render the person autonomous and responsible for his/her career development and personal growth.

Target group

It must be made clear that the 'centres' were not created specifically to take care of the unwaged. They are just as much for those people who, while still in employment, find they are on the point of losing their jobs or who want to change jobs. Clients generally come from semi-skilled backgrounds and many are women according to a survey done in 1987 and a study done by the present author in 1991. However, there are also some technicians and people from executive levels.

People are motivated to have a 'bilan' because they need an assessment before a career change for personal development, retraining or simply because they are threatened with unemployment and, therefore, feel they will become marginalized.

The 'centres de bilans'

It was soon agreed that these assessment and guidance procedures should take place in a location which was 'neutral' – ie, not linked either to educational institutions or employment departments – though they may use company or institutional infrastructures.

In 1986/87, 15 pilot centres were set up and various specialists were brought together to assist people seeking a 'bilan'. The positive results of the experiment led to 32 more centres being set up in 1988. In 1991, there were 82 fully operational centres throughout France.

The members of the inter-disciplinary team have been released either partly or fully by their original institutions and must have a good grounding in local socio-economic environments.

Financing

The average cost per hour is about £10 to £15 for an average of 15 to 18 hours of 'bilan'. Clients, despite being volunteers, rarely pay for their 'bilan'. In the case of employed adults sent by their companies, the employers bear the cost. However, as confidentiality is a critical aspect, the 'moral' contract made beforehand lays down which results are available to the employer. For the unemployed, the cost is borne by the state or local governments through job searching agencies.

Some key principles

There are some professional ethics mentioned in official texts:

- The client is to come to do a 'bilan' on his/her own free will. No employer or institution must exert any pressure on him/her. The whole procedure requires strict personal involvement and has to do with the very identity of the person.
- Confidentiality is a fundamental aspect.
- The whole process is negotiated with the client and everything – assessment, results, etc – must be totally clear.
- The detailed results belong strictly to the client. When an employer pays for the 'bilan', what is to be revealed to the company is discussed with the client and subject to his approval.

The reaction of employers

In France, according to the law, firms and companies employing more than ten people have to devote a minimum of 1.2 per cent of the wage bill to training their staff. Some big industries and companies allocate as much as seven or eight per cent of the wage bill to this. Employers consider, therefore, that training is an investment and closely linked to the human resources development policy. The 'bilan' enables them to detect potential in their staff and helps them to set up programmes of reskilling or redeveloping personnel to reduce job losses and cope with market and technological changes.

Problems and issues

Common questions raised are about how to do justice to the individual and how to provide more flexible and individualistic forms of assessment. It was clear that traditional methods of assessment, especially norm-referenced testing, were not appropriate.

However, if there are no strict rules applied and no formal testing, how far can such an assessment system embody the validity, reliability and credibility that give an assurance of quality and, hence, national acceptability? Debates on this are still going on.

Other issues raised concern organization, management and resourcing. One of these is the question of how institutions, through their staff, could work together in new and more intensive partnership, fostering joint staff development.

Because the main partners in this venture are education and employment/vocational training providers, there are conflicts relating to the roles of the staff involved. Though a functional framework for professional activity has been found, it still requires a lot of goodwill from people who have long ignored one another and must find common ground to function.

Also, there are questions about:

- how to guarantee complete neutrality and independence for the centre – usually centres are autonomous as far as budget is concerned, but make use of the existing infrastructures of either educational or employment and vocational training institutions;
- who should manage the centre; and
- which institution will provide a technical director and who will choose the members of the team, which should be both inter-institutional and interdisciplinary, so as to maintain a sound balance between the institutional partners involved.

The state subsidizes the different centres, but which institution enables an individual centre to go on operating while waiting for subsidies which do not arrive in time?

As one of the staff from a centre put it: 'To translate inter-institutional functioning into effective partnership in this case, there is a need for strong will coupled with a good sense of humour.'

A fundamental question, in my view, concerned the target group addressed by the CIBCs. Initially, the centres have not been created to cater for the needs of the unemployed or the unskilled. Nor are they aimed at target groups with special needs, although there are some specific initiatives designed for them. The CIBCs, as already mentioned, look after people who are on the way to being marginalized because of technological changes, redeployment of personnel or corporate economic difficulties. They represent an answer to a particular social demand. What happens if the unskilled, the already marginalized or members of ethnic minorities seek access to a 'bilan'? Are they to be refused a service which would help, besides other things, to build their self-confidence and assertiveness?

The official text about the CIBCs points out that they are 'to receive anybody who volunteers, whether they are employed or not, young or adult' (Circular No. 36, 14 March 1986). But the 'bilan' process does require from the client a certain level of education. So, what happens if he/she does not possess the intellectual ability to enable him/her to benefit from it? Can a team or a counsellor ethically refuse to accept the request of such a person?

In the context of economic depression and unexpected and unprepared changes, the word 'bilan' has come to be associated with hope for people in need. Consequently there are many examples of 'head hunting' and private human resources consultancy firms offering some kind of 'bilan' at a very high cost to individuals who would not have access to a CIBC. A recent study of the initiatives of 'bilan' in the south of France by the present author highlighted many problems related to assessment and work ethics of professionals involved. The question, therefore, is where to exert a control for quality assurance in this case.

The 'bilan' empowers people to use their identified competences as a negotiating tool because, during the process, they have learnt what potential they have. A recent agreement between the state and social partners in France (3 July 1991) gives all employees the right to a 'bilan' paid partly by his employer. But, if we consider the purpose of the 'bilan', we are then assuming that such negotiation is possible in the work-place. Isn't this not only optimistic but somewhat idealistic?

Employees who have gone through the process usually come out mastering their potential and with a new self-image. Yet, they will most likely go back to a work environment where nothing has changed or even been questioned. Can firms and industries be expected to review long-established habits and salary scales set up after collective bargaining agreements? Then, will not the 'empowered' employee just end up feeling frustrated? Or, has the 'bilan' heralded a social revolution in the work-place? Only time can tell.

18 Community Involvement and Development Course

CONTRIBUTORS: JOL MISKIN AND KATH WOODWARD, WORKERS' EDUCATIONAL ASSOCIATION, YORKSHIRE SOUTH DISTRICT

The 'Social and Community Skills Vocational Training Project' was launched by the WEA Yorkshire South District in the Autumn of 1987. Funding was obtained from the ESF enabling the appointment of a part-time worker whose task it was to organize, recruit and direct a special course for the long-term unemployed. The course quickly lost the above title to became the rather more manageable 'Community Involvement and Development Course'.

Jol Miskin was appointed as Tutor/Organizer and, in the best traditions of the WEA, and adult education more generally, was asked, with little support, to get on with the job. The then District Secretary, Ray Fisher, seemed to have his reservations about the whole thing which he stated honestly in his introduction to a report Jol Miskin wrote following the first and, in the event, extremely successful course which ran from April to June 1988. 'I was not completely happy about making the appointment,' Ray wrote, 'and did so with some hesitation. Recruiting unemployed people to WEA courses is one of the most difficult and daunting tasks facing adult educators at the present time. To get a commitment to a five or ten meeting course is difficult but asking for a positive and demanding commitment to a two-day weekly course of three months duration was just about asking for the moon. I was convinced that some sort of reduced or shortened version might be a possibility but that a three-month course was a non-starter.'

Ted Hartley, responsible for the Rotherham Metropolitan Borough Council's bid for ESF money, was more confident in the ability of the WEA to deliver. 'I had as an objective,' he said, 'the goal of recreating, in some part, the historic tutorial class. I feel that the intensive 200-hour course has moved the WEA nearer to achieving that type of sustained, methodical and enriching process.'

It did work and, indeed, continues to do so three years on. So, what have been the ingredients for success? What have been the problems and issues? And, finally, what scope is there for development?

The official project title was 'tailor-made' for an ESF application. An emphasis had to be placed on hard vocational training, incorporating information technology in the curriculum. Furthermore, the proposal envisaged recruitment of 30 students, a high percentage of whom would go into employment on completing the course. This was somewhat fantastical and we assume that being a little 'economical with the truth' in form filling remains the norm for most applications. What we mean by that is this. Firstly, you cannot just expect to recruit 30 long-term unemployed adults to a fairly long course by simply advertising it. You have to go out and win them over to the idea of 'returning to learn'. Secondly, recruiting 30 students to one course is both extremely difficult and, even if you succeeded, educationally unsound – 30 students in the class! Lastly, anyone with the slightest knowledge of South Yorkshire in the 1980s and 1990s, would know that obtaining employment, even with good qualifications, is very difficult. Unemployment has been very high for over a decade. So, for 30 students to get jobs after a 200-hour WEA course would have been nothing less than miraculous. Everyone, including the funders, knew this and yet they all seemed to go along with the charade.

However, helping people onto the ladder to a job or to FE/HE was quite feasible and so this became our aim. To that was added the aim of enabling students to play a more confident and effective role in the development of voluntary and community organizations. In short, our course was to offer a variety of 'progression' routes, but underpinning it would be education with a 'social purpose'.

The 200-hour course, more recently extended to 250 hours, was not designed exclusively by professionals or on the basis of EC directives, though both contributed to the curriculum. Rather, Jol Miskin, with the support of the WEA District, embarked on a six-month period of developing the project before piloting a first course. What form did this take?

(1) Making contact with a variety of groups in the community to discuss the project aims and to listen to education and training demands. Over 100 such meetings took place, including discussions with trade unions, ethnic minority organizations, community groups, etc).

(2) Putting on 'taster' courses, some of which were a result of express demand, others being 'tutor led' on the basis of the listening exercise outlined above.
(3) Arranging some 'community forums' to address current issues facing working-class families in Rotherham (eg, privatization and the 'benefits' system).
(4) Organizing publicity in the form of leaflets and newspaper and radio coverage for the project.

This development work proved absolutely crucial to the final delivery of the course and confirmed that winning back to education adults who have largely been failed by the education system first time round is a sensitive job which must be done methodically and efficiently and with adequate resources. There are no short cuts.

Our approach proved more than effective. A final publicity leaflet, to which a pre-paid postcard was attached, brought in 32 applicants for the course. Maybe we could get 30 students!

In the event, a process of informal discussion with each person expressing interest in the course reduced the final intake to 18. Further courses have recruited none, 14 and 15 respectively.

The students have been predominantly working-class and most have lacked any formal qualifications. Overall, there have been roughly equal numbers of men and women. In terms of age, students have been quite evenly balanced throughout the 20s, 30s, 40s and 50s. The number of black students has been low – four in 1990 and one in 1991.

The curriculum

The pilot course incorporated four key learning themes:

(1) the political system and 'power';
(2) neighbourhood and community;
(3) the welfare state; and
(4) information technology for the community.

Kath Woodward, who took over the Project in 1990, also introduced her own mark on the course and further developed the curriculum to include personal and work-related action plans, development of competences and transferable skills and some work experience.

While these have been the themes of the courses, underpinning the entire process from start to finish was the building up of personal confidence and self-esteem along with the development of a variety of organizational and study skills. Learning methods centred on group work but also included short lectures, the use of video material, study visits (including the Houses of Parliament), guest speakers and project work.

Students build up their own 'records of achievement', in the form of a file, as the course proceeded, which means that on completing the course they have a file to be proud of, incorporating a variety of personal assignments, a project report, notes and handouts and tutor's comments. All students with over 80 per cent attendance records received WEA Certificates of Attendance and our evidence is that the course aims have been achieved for a high percentage of the intake.

Students have, in a number of cases, moved on to further educational opportunities. Others, albeit a small percentage, have obtained employment almost immediately and we are confident that others will do so in time, having taken the first step of participating in this course. Finally, most have become more involved in the community as 'active citizens'; 'passive subjecthood' is no longer for them, if it was in the first place.

The resources for course development were never as good as the first time round. This is a shame and the effects are confirmed in the recruitment figures. Nonetheless, the course has remained remarkably successful and has clearly found a niche locally. Adaptation and change have occurred partly due to different tutor input along with individual tutor's respective perspectives and approaches to the work, but also due to ongoing student evaluation of the learning process. This has taken the following forms:

- individual tutorials;
- group discussion and feedback;
- student committee;
- student 'personal profiles' and 'diaries';
- informal discussion; and
- questionnaires.

A combination of these evaluation methods has helped illuminate the process, thereby enabling adaptation of the course in line with student demand. The course, in short, has sought to be student-centred, highly participatory and democratic.

What then, are the key aspects making for such effective provision?

(1) The course is not narrowly vocational. It is not about providing specific skills so as to slot students into specific types of work. Rather, it extends the student's knowledge and powers of ana-

lysis. It helps the student to understand more thoroughly the world in which she/he lives and, thereby, to make decisions as to her/his future and that of the community. It offers a broad based learning opportunity and also emphasizes a variety of specific forms of study, along with providing organization and 'new technology' skills that together enable students to move on to new and unconsidered areas. This opens up employment opportunities and develops a range of transferable skills, for employment, work in the community and in personal life. We contend that confident adults, alive to social and political issues and concerned for their fellow citizens, able to develop critical analytical skills which lead to an active questioning of social values, bolster and improve the world we live in and the work-place. Our approach leads to discussions about the kind of society, communities and work-places which we want. All these issues are raised in the course and students are encouraged to ask questions and engage in critical debate.

(2) It is well resourced though, needless to say, there is always scope for improvement. Access, in the form of childcare, free travel and free provision, is crucial and has ensured the recruitment of those who otherwise would remain firmly at the other side of a closed door to educational advancement. 'Access' also incorporates a pleasant environment in which to study and friendly supportive staff, both tutors and other workers in the WEA District office.

(3) Individual tutorial support has also been of the essence as students bring a variety of problems and worries to the course, including personal ones which need to be sensitively talked through and addressed. Advice, guidance and counselling are recognized as being integral to work of this nature and need proper resourcing if real progress is to occur.

Overall, the course has been effective because students have been convinced that taking what is often a very big step and participating will not mean having to expose their weaknesses or limitations but, rather, will enable them to tap their strengths and hidden potential. Our approach ensures that students succeed. It recognizes the innate abilities of working-class adults failed by our educational system. It is this belief which is the driving force behind the course's success and effectiveness. We are sure that this will continue to be so. Much more of the same is needed if this or any future Government is really serious about creating a 'healthy' economy or a 'well-trained workforce'. It really is a question of citizenship against subjecthood, of whole confident adults as opposed to limited and restricted ones.

19 Joint Initiatives of Industry and Trades Unions in Adult Education

SPEECH TO THE EUROPEAN CONGRESS IN ST GALLEN, SEPTEMBER 1991
CONTRIBUTOR: JACQUI BUFTON, GLOUCESTERSHIRE LOCAL EDUCATION AUTHORITY

In recent years in the UK there have been a number of joint initiatives undertaken in the field of adult education and training by both sides of industry – employers and the trades unions.

Although joint partnerships and approaches have to a certain extent been discouraged at Government level, at company level there appears to be a continuing realization of the need to work together to solve common problems, such as the need to develop and sustain a skilled work-force.

Changes in working practices are demanded by companies wishing to modernize and become more competitive, and it is often at this point that there is an opportunity for employers and trades unions to negotiate a commitment to education and training for the workers, as part of the wage and salary package which is agreed on.

In some cases trades unions have been quick to seize these opportunities and are actively encouraging their officers to press for management support for education and training as part of the package on offer to their members. They recognize the importance of education and training in sustaining their members' employability and job mobility in what is an increasingly changeable and competitive industrial climate. In fact, the largest UK union, the Transport and General Workers Union, has produced a package for its members called 'Negotiating Training at the Workshop'.

For their part, the employers have taken the view that encouraging and supporting the education and training of their employees is the best way to ensure a continuing supply of skilled and flexible workers. The learning process itself is regarded as giving workers the potential to respond rapidly to the inevitable changes in working practices that employers will require. Learning means looking at things differently and being prepared to accept change – positive attitudes that employers need from their work-forces today.

The UK's employer organization, the Confederation of British Industry (CBI), has long recognized that the key to improving Britain's competitiveness is to encourage people to participate in education and training after they have completed their 11 years of compulsory education. Britain has one of the lowest participation rates in post-compulsory education of all the Organization for Economic Co-operation and Development (OECD) countries. The CBI is encouraging its members to invest in their workers' potential through devising company policies which will utilize people effectively. Sir Bryan Nicholson, Chairman and Chief Executive of the Post Office and Chairman of the CBI's education and training committee, has emphasized the point in a recent statement: 'Employers,' he said, 'will need to invest more in people, to concentrate on updating skills levels and broadening skills to enhance flexibility and responsiveness.'

Jointly negotiated initiatives in the field of education and training have the potential to improve the work-force which both sides of British industry recognize is needed both now and in the future. This affords a recognition of the role of adult continuing education in the work-place and offers an important boost for those who have been supporting and providing adult education services.

Let me now describe in detail The Ford UK joint Employer/Trades Union Initiative: 'The Employee Development and Assistance Programme (EDAP)'.

The Programme is designed to provide education and training opportunities for Ford employees. Its key characteristics are that:

- any education and training supported by the programme is carried out in the employee's own time (ie, non-work time);
- only education and training which is non-job-related is supported by the programme; and
- the programme is open to all employees.

The origins of the scheme

The scheme originated in the USA in the early 1980s and began as a response to large-scale redundancies in the Ford factories there. The UK scheme is to a certain extent modelled on the American experience, but with a key difference – the UK programme rose out of the 1987 round of wage negotiations and was seen by the unions as part of a package of improved conditions and benefits and is not, therefore, a response to threatened redundancies.

The role of Ruskin College

Ruskin College, Oxford, is a college for adult students and is supported by the Trades Union movement. It has a long-established (21 years) Trades Union Research Unit which has always worked with the Ford unions on the preparation of wage claims. Ruskin, therefore, worked with the unions in formulating the EDAP scheme as part of the November 1987 wage negotiations. The Trades Union Research Unit has continued to be closely involved in the programme and devised and administered the initial 'training needs questionnaire', which went out to every Ford employee. This is probably one of the largest surveys of adult attitudes to continuing education and training ever undertaken and had a response rate of 13 per cent. The information gained from the survey was used to help to plan the programme. This also proved to be an important way of publicizing the programme and introducing the work-force to its aims.

How it works

Every Ford employee is entitled to a grant of up to £200 per annum to undertake non-job related education and training.

For an employee wishing to take advantage of EDAP, the first stage is to make contact with the 'local education adviser' who is employed by the programme to work in the factory and his/her job is:

- to provide information and advice on the education and training opportunities available;
- to organize courses on factory premises; and
- to negotiate with potential providers on behalf of Ford employees.

The local education adviser works for the local EDAP committee but, in order to demonstrate the independence of the programme from Ford management, the advisers are employed by the Polytechnic of East London. Those devising the EDAP programme recognized that the provision of information, advice and guidance was crucial in helping employees to gain access to the education and training opportunities available to them through the programme.

Once contact has been made with the local education adviser, the employee completes an EDAP application form which then goes forward to the 'local EDAP committee'. Every factory, however small, has its own EDAP committee and these committees are a crucial illustration of the philosophy of EDAP – that it is a joint management/trade union initiative. They are therefore made up of representatives of management and hourly paid and salaried trade union representatives. The local committees also demonstrate the other agenda of the programme, which is to try and bring about an improvement in industrial relations through the process of working together on something positive. The management of EDAP is described as tripartite, reflecting the involvement of the two types of unions, hourly paid and salaried, and Ford management. The members of the committee are responsible for considering each application and deciding whether to support it.

Funding: The programme is funded by the company through a per capita contribution for every employee on the company's payroll. This currently stands at nearly £50, and it is raised pro-rata whenever there are pay agreements. This money is paid into the EDAP fund and the money is disbursed to support the work of the programme. The fund currently stands at over £2 million. Once the money to pay the local education advisers is taken out, the amount left is distributed to the local EDAP committees, again on a per capita basis, so that the bigger the factory the larger the budget available to the local committee. Committees make their own rules about what they will fund under the EDAP scheme.

Although a recent Ruskin evaluation of EDAP stated that the programme made no claim on public funding, this is not entirely true in that Ford employees who use their EDAP allocation to sign up for local education authority provision are in fact being indirectly subsidized through money given by the Government to local authorities for the adult education services. The Government, through the Employment Department, is also contributing to the funding of the salaries of the education advisers through a grant of £150,000 per annum for the first two years of the programme's operation.

What it has achieved

Numbers taking part: The figures given out by Ford are that 45 per cent of the work-force has enrolled on courses – 20,028 employees. These figures though should be treated with some caution as they will relate to the number of applications under the EDAP programme. A number of employees make more than one application during the course of the year. The recording system in use counts applications rather than individuals. Nevertheless, the programme has clearly attracted an impressive number of people into adult education.

Responding to the needs of shift workers: Most Ford plants operate shift systems. To meet the needs of shift workers, a great effort is made to provide courses on the factory premises. This has undoubtedly increased participation because it enables workers to undertake their study either immediately before or after their work shift and has removed the barrier of having to find the motivation to go to another venue after returning home from work. For shift workers in particular, whose pattern of work might change on a two- or three-week basis, this has enabled them to keep up with their studies.

The building of links with local providers of adult education: Strong links have been built by some EDAP education advisers with the local further education college and with adult education centres and nationally with the Open University. There are also the beginnings of links with Open College Federations, where they exist, and these are providing accreditation opportunities for workers who are undertaking courses which have been specially laid on for Ford workers on factory premises. The adult education world in the UK has been anxious to respond positively to EDAP and many providers have been able to demonstrate great flexibility in terms of the mode and time of the delivery of their programme.

Choices being made

The latest information about the choices people are making shows that a quarter (4,500) are studying foreign languages. French and Spanish are being studied but the most popular language is German, perhaps because of Ford UK's connection with German factories. Craft skills – eg, bricklaying, carpentry, etc – are popular and computer skills attract a large number of students, with some local committees spending part of the budget to buy computer equipment for EDAP students to use. All possible areas of adult education study are covered, including, through the development of links with local higher education providers, the provision of degree courses on factory premises. Health and fitness pursuits are also funded by the programme and are seen by both management and unions as encouraging workers to adopt a healthier lifestyle. There is some evidence to suggest that blue collar workers are choosing to study vocational subjects rather than the academic or leisure ones and are perhaps seeing EDAP as offering the opportunity to undertake some retraining in preparation for further employment.

Effects on work-place morale

One of the major benefits perceived by Ford management of the way in which EDAP is run, is that it would provide a vehicle through which industrial relations could be improved. The joint management of the programme and the fact that the provision of courses on the factory premises might bring together a group of factory-floor workers, the managing director and clerical staff all learning a language together would allow for some changes in perceptions and attitudes on both sides of the industry. It is perhaps too early to say whether EDAP can demonstrate an improvement to industrial relations but, anecdotally, there is some evidence to suggest that in certain factories, facing the management issues of EDAP together has given both sides a greater understanding of each other. In fact, EDAP has now been moved within Ford from being the responsibility of the education and training function to that of industrial relations.

Its future

EDAP has demonstrated that there is a demand from all sections of the Ford work-force for adult education of a variety of kinds. Ford UK and the unions entered into the biennial wage negotiations in November 1991 and the continuation of EDAP was on the agenda. It was supported by both sides of the industry. It therefore looks as if it is here to stay. Much has been achieved already by the programme.

The next phase will need to explore ways of encouraging those in the work-force who have not yet taken advantage of what is on offer, to do so. Although by most standards, EDAP has a large budget, it is still a cash-limited programme. As with all cash-limited programmes, some targeting therefore needs to take place and the local committees will have a responsibility to ensure that everyone in the work-force has access to the opportunities the programme affords and that it is not just open to those who have already participated. Very few opportunities have been offered through EDAP to those who might have numeracy and literacy difficulties, whereas much publicity has been given to degree-course provision. This is perhaps inevitable, but the purpose of EDAP might be better served if the emphasis was the other way round. Because different committees are agreeing to support different opportunities, there is an inequality in what is supported throughout the country. This is the price of the devolution of responsibilities for the programme and yet it is something which those responsible for the development of EDAP will need to consider. Devolution and pro-rata funding also means that smaller factories have a narrower curriculum on offer on the factory premises, which is a particular disadvantage for shift workers.

As in everything, improvements can always be made but, so far, EDAP has clearly demonstrated that with the right kind of promotion and support, adult education will be taken up by large numbers in the work-force. The predicted response rate for the first year of EDAP (based on the American experience) would be that five per cent of the work-force would participate in the programme. EDAP, with an estimated 45 per cent participation rate two years into the programme is, by any standards, a remarkably successful example of what can be achieved through joint initiatives.

Other related initiatives

Other large companies have also set up joint initiatives, usually through the wage and conditions of service negotiations. ICI (Imperial Chemical Industries) has recently introduced an employee development programme for their manual workers as part of a deal directed at a radical revision of working practices. The ICI programme will differ from the Ford EDAP programme in that the company will allow workers to take courses in company time.

Other large companies such as Lucas and Rover have also introduced similar programmes. Rover set up 'The Rover Learning Business' (see Chapter 20) last year which offers study programmes (self-development), and is complemented by the Rover Employees Assisted Learning Scheme which provides up to £100 in grants for learning courses. 20,000 employees (half the work-force) have taken advantage of the study programmes offered. Rover sees it as an important way of encouraging employees to realize the positive contribution which they can make to the business.

However, it is not only the large companies which are taking such initiatives. Through the support of the National Institute of Adult and Continuing Education's REPLAN programme, a small company in Sunderland is, in partnership with its local Trades Union Council, examining the attitudes and motivation to learning among the work-force and the barriers which exist to the participation in continuous learning. The purpose of this joint initiative is to develop a continuous learning culture within the enterprise. This is generating a demand for adult education and training which is being met through a partnership between education and training providers.

Nationally, organizations involved in supporting adult education and training are working to encourage joint initiatives. In particular, 'Workbase' – a national body supported by the TUC and the CBI –

has, for the last ten years, been working jointly with managements and unions to provide basic communication skills for manual workers. Workbase is often brought in at the point of technological change, and is paid for by the company, with training taking place in work time. Training is often delivered in partnership with local adult education providers.

ALBSU, the national Adult Literacy and Basic Skills Unit, is currently receiving Government funding to undertake a number of local surveys of employers' needs in terms of the basic skills of their work-forces. This initiative is designed to support employers in their drive to increase the skill levels of their employees and to encourage increased participation in education and training at all levels.

NIACE too has its own initiative – 'People, Learning and Jobs' – and is supporting the idea of paid educational leave in its recent paper *Towards a Learning Workforce: Adult Learners in Industry*. This suggests that there should be legislative entitlement for all 16- to 65-year-olds to paid learning time of 30 hours per year, plus 30 hours per year of unpaid study leave.

Central Government support for training opportunities for adults has recently resulted in new legislation which enables anyone paying for vocational training to be entitled to a tax rebate on the fee charged. The focus of some proposed legislation is described in the White Paper on Further and Higher Education, which also clearly supports vocational provision. Europe, too, is beginning to focus on vocational education through the new European Commission programme, FORCE, which is supporting vocational training opportunities for adults and is promoting a joint industry and trade union approach.

It remains to be seen whether a broad educational programme such as Ford's will be the model for the future or whether support for such an educational programme within companies will give way to a more tightly controlled specifically vocational and work-related approach. It is perhaps interesting to note that the responsibility for the EDAP programme now lies with the industrial relations section in Ford, rather than in the education and training section, which is where it began. This suggests a perception of adult education as a contributor to joint working relationships rather than a contributor to the upskilling of the work-force.

However, adult education in the UK is something which both sides of industry are now taking seriously, even if it is for a variety of reasons, and this must be welcomed. This development will inevitably mean changes in the way adult education is expected to be delivered – and changes too in the profile of the participants. The

PART III: RESPONDING TO THE CHALLENGE

partnership of industry and trades unions and these joint initiatives offer a whole new territory of working in adult education. It is the business of those working in adult education to make sure that they are in a position to respond to these developments, creatively and enthusiastically, for the benefit of all.

20 The Rover Learning Business

WRITTEN BY JENNY HUNT FROM ORIGINAL MATERIAL PROVIDED BY
FRED COULTER OF THE ROVER LEARNING BUSINESS

Rover is one of a growing number of large companies in the UK which have adopted an entirely new philosophy towards the training and development of employees. It has developed an initiative called the Rover Learning Business (RLB), which has been set up to achieve the following key objectives:

- to emphasize the view that people are the company's greatest asset;
- to unlock and recognize employee talents and to make better use of these talents;
- to convince its employees that the company's commitment to every individual has increased;
- to gain, both internally and externally, recognition for employee achievements though internal and external accrediting bodies; and
- to improve the competitive edge of the company.

It is an exciting and new idea because of the emphasis the company is placing on actual learning and individual needs. It is hoped that the use of the word 'training' will gradually disappear because it implies something which is done *to* people whereas learning is something that people do for themselves.

The new RLB is a business within a business and was launched in May 1990 with a budget of around £30 million. It has a company structure like any other business with a Managing Director and a Board, known as the Board of Governors. The Business's customers are the Rover Group's 40,000 employees, who will be helped to improve their own knowledge and skills in line with the business needs of the Rover Group, no matter what their job, what their grade or what their educational background.

A detailed survey of their employees provided Rover with real proof that there was a great deal of untapped potential held in its employees and, by encouraging people to develop this potential, it was felt that both the individuals and the Rover Group could benefit greatly. Since the introduction of RLB, prospects for employees have looked very good. The company has agreed to pay for all training that is relevant to its employees' job of work but, in addition to this, RLB has introduced a 'product' called REAL (Rover Employees Assisted Learning) which offers up to £100 per annum to each employee for a learning experience that does not have to be directly related to the job of work. For example, someone on production may want to learn French or computing and, although the course may not be directly allied to the job, the company feels that the person will be better for having taken the course. It is this valuing of people and their needs which makes the scheme so different, so desirable and in line with the current CBI 'Investors in People' philosophy.

Because the company wants, in particular, to make learning opportunities more accessible for its employees, it is planning to provide courses on site at the end of working shifts, so that people don't have to go home and then have to make a great effort to get out to college in the evening. This way, they will be able to finish work, have a coffee at the training centre, do the course and then go home. This approach, the company hopes, will encourage greater, more regular participation on courses and make them less of a chore.

The development of such courses requires close collaboration between Rover and local colleges and training providers. This partnership approach to the development of opportunities for people within the company is very important because it is a major step in breaking down workers' traditional attitudes about colleges being 'not places for people like us'. Gradually, Rover is introducing a climate of continuous improvement through the provision of opportunities for continuous learning, so it is not just the company that benefits, but the employees too will have their efforts rewarded. To ensure this, Rover has arranged for courses to be accredited externally by bodies such as BTEC and the National Council for Vocational Qualifications or, where programmes do not fit comfortably with the criteria of an outside body, there are rigorous internal standards marked by certificates of competence. It is also possible that successful completion of programmes may be marked with an

RLB Certificate that is endorsed by an outside body. In this way, quality is guaranteed and employees can gain nationally recognized qualifications.

Each employee has a record of achievement and every time a programme is completed, or a skill acquired, it will be formally entered on this personalized record.

When Rover employees decide to become part of the RLB, they are first given a personal development file which is their own property and which provides a lifelong record of their personal and work-related learning, development and achievement. It is also a visible demonstration of the company's declared intention to support and encourage all its employees to develop through learning.

The employees then discuss, with their manager, achievements to date and aspirations for the future in terms of both job satisfaction and personal development opportunities. During this interview, employees are encouraged to think about themselves, their jobs and where they feel the need to develop. The manager then helps to clarify development needs, analyses future activities in terms of skills, provides guidance on available learning opportunities and identifies the best methods of achieving agreed aims to suit individual needs and eventually draws up a personal development plan.

One of the options which they may choose to try is one of the 'products' or courses actually produced by RLB. Subject areas available include 'health education', 'learning to learn', 'environment and green issues', 'Europe 1992' and many others.

There is also a diary section in the file to help the employees to schedule their learning.

Having attended this initial interview, the employees then become responsible for their own learning and take ownership of their own development plan.

Since the RLB was established its success has been measured by the following statistics:

- 20 per cent of employees were working on learning programmes in their own time and related to the REAL product by the end of 1991;
- 50 per cent of employees had personal development files by the end of 1991;
- 1 per cent of employees completed Rover manager-approved programmes by the end of 1991; and
- 6 per cent of the annual payroll has been invested in employee learning and development.

PART III: RESPONDING TO THE CHALLENGE

All of this represents a real commitment by Rover to the individual development of its employees. Also, if it continues to be successful, it will, over the years, gradually transform the employees' attitude to learning and training and create an exciting new learning culture. The company's dedication to this new kind of philosophy for industry is demonstrated in this statement by Sir Graham Day, Chairman of the Rover Group and the RLB:

> Growth by any person, plant or living thing must come from within, and it is this principle which has brought the RLB into being. The primary job of the Business is to bring about growth by releasing the talents and energy which already exist within. Rover spends something like £30 million a year on education and training and it is my belief that this should be targeted to identify, encourage and develop the abilities of individual employees, no matter where they work.
>
> Building on Rover's success in developing the Business and its products, we now aim to provide an environment of continuous learning, an environment in which improvement flows out of the hearts and minds of our employees at all levels. This is what the RLB is all about, and I encourage you to take advantage of all that it offers.

21 Meeting the Needs of Industry

THE WESTMINSTER COLLEGE EURO-CONSULTANTS COURSE
CONTRIBUTOR: GERRY SMITH, LONDON BOROUGH OF
WANDSWORTH

Unemployed education and human resource management

For many years, the education and training of unwaged adults has been seen as a necessary function of a humane society to get the unemployed back into stable employment. In reality this has often resulted in a series of programmes which, at best, provided new opportunities for a few unemployed people to explore new career paths, or, at worst, filled in time and enabled unemployment to be recycled, but provided no real employment growth within the geographical area.

In the recession of the early 1980s, the main groups affected by unemployment were the unskilled and semi-skilled. At that time it was the restructuring of viable components of markets which determined unemployment levels. The decline in manufacturing (including coal and steel), for example, meant that industrial regions of the north were particularly badly hit with new levels of structural unemployment. The early 1990s recession, however, has had a very different impact. The closures of businesses and industries are not so much market specific, but are more based on changes in the way businesses are managed and financed within Britain. The once top-heavy middle management of many companies is now on the decline as companies rationalize person power. The boom in small and medium-sized enterprises in the 1980s and the subsequent closure of many of those by the 1990s led to the emergence of a new group of unemployed people in all areas of Britain and other EC states.

PART III: RESPONDING TO THE CHALLENGE

This new group of unemployed adults was highly educated, highly motivated but still suffered much of the psychological depression affecting other groups, such as lack of confidence and feelings of rejection. Any course programme developed for these people therefore needed to include guidance, counselling and student support systems.

Identifying industries' needs

As 1992 was drawing closer it became apparent that existing businesses in many European countries had in no way prepared themselves for the legislative change that would affect their business in the near future. Within a recession, all energies and resources are focused on survival rather than development and planning. It was the combination of these elements which led to the development of the Euro-Consultants Training Programme.

The key concept was to provide unemployed graduates and managers with specific European knowledge and networks to provide human resources for local companies.

Course credibility (a new Euro-course)

Many courses in Britain offer courses with a European dimension. The difference with the Euro-Consultants course is that it was developed in partnership by providers in a number of different countries. Initially the project started as a short course run between Denmark and The Netherlands. The course is now one academic year in length and spans Belgium, Britain, The Netherlands and Denmark. France and Spain will be joining the programme in 1992.

The programme is effectively one course, with over 100 students, delivered in each partner country.

Students from all countries have a number of opportunities to meet throughout the year either in host countries or in the annual one-week visit to Brussels. Work experience takes place in host countries or even farther afield. For example Dutch students are currently working in Czechoslovakia supporting trade links with small business in EC countries.

Students are given a core studies element which provides them with a base knowledge of European development. Then they specialize in a number of key professional business areas. Each student has to complete a number of projects on a specialist area. Students are given facilities for communications such as fax machines, international telephone lines, etc, and their projects must include a truly international element. Projects are assessed partly on research through national data but also by the comparisons drawn to real business practices in other countries. For example, a student who wishes to explore Dutch marketing techniques will not only read the relevant books but also ask through the network how Dutch marketing techniques differ from English marketing techniques.

Guest lecturers have included Members of the European Parliament and the London director of the EC.

Student profiles

Unemployment has a similar effect on all people in the community, including those from professional backgrounds. The fact that they have 'done the right thing' – gone to college, shown dedication, taken out a mortgage, etc – also means they feel rejected and incompetent with their own skills. The course therefore needed to provide a structural support system which enables students to understand their unemployment and how to manage their own change of circumstance and careers. Guidance and counselling is therefore integral to the programme. A student who turns up in a suit, with a Samsonite briefcase, may be in the final stages of becoming homeless through mortgage default.

The students, while learning about company business plans, have also therefore got to produce their own personal development plans.

Course management

INTERNATIONAL STEERING GROUP
|
NATIONAL STEERING GROUP
(Education/College/Economic Development Office)
|
NATIONAL COURSE DIRECTOR
|
COURSE TEACHING TEAM
|
STUDENTS
|
FEEDBACK

The course in Wandsworth is a collaborative venture between The Further Education Department and The Economic Development Office. This is replicated in all partner countries. Only by having co-ownership can both the students and local business be adequately served.

Funding

The key source of funding for the programme has been through the European Social Fund. This has also helped the programme to develop a number of valuable links to various departments within the European Commission.

At delivery level it became apparent that the course was relatively expensive in comparison with other college courses. Foreign visits and international communications are not cheap and therefore it was imperative that these burdens should not in any way fall on the student.

At the same time students had to be able to reorganize their home lives to allow them to travel when required. Childcare was also, therefore, an important factor to be considered.

International issues

International Steering Group meetings for the course always provide stimulating debate. Course content and delivery methods vary from country to country and have to be reconciled. For instance, equal opportunities policy may be a key factor in Britain but not high profile in The Netherlands. Similarly, languages are not an issue in The Netherlands, where most of the students speak a number of languages. This is not the case in Britain.

Issues for a further education college

To develop and deliver a new international course for higher level students, a number of key issues have to be addressed:

- adult students with previous management experience will expect good course management;
- adult learners will not accept off-the-shelf lessons which have been used for many years on other 16–19 courses;
- elements of the syllabus must be negotiable;
- the course must utilize the existing skills and knowledge of the students;
- staff must be prepared to travel and communicate internationally;
- financial resources need to suit the course requirements;
- staff development time needs to be allocated to ensure the course reflects the changing face of European business;
- regular feedback from students is essential to ensure they attend;
- course team meetings are required at college and international level;

- staff need to be trained in the learning needs of adults;
- recruitment needs to reflect the local community (for example advertising in the ethnic minority press) and;
- a course of this nature needs to have a steering group to support the staff involved and to monitor and evaluate performance delivery and finance.

Outcomes for students

- To take up self-employment in European consultancy. (This may vary from giving expert advice to individual businesses to advising voluntary sector organizations on European funding.)
- To take up employment in local firms which decide they have a need for a European specialist.

Conclusions

Setting up a transnational course of this nature can be a very difficult task. However, the benefits to both students and local enterprise are almost unmeasurable.

For many years we have discussed developing courses to suit local skills shortages. This in itself becomes self-perpetuating as we fail to see the wood for the trees. Often we have just given more students more qualifications for the same jobs. This programme deals with real economic needs in a Europe that is changing dramatically. Students on this course, even those disadvantaged by age or race, soon become some of the very few people in an area with unique skills. This can only benefit both the students and the local economy.

PART IV

Summary and Recommendations

22 Analysing the Past, Shaping the Future

'IF I COULD DO IT ALL AGAIN, I WOULD START WITH EDUCATION'
(JEAN MONNET, 1954)

Analysing the past

Britain's failure to train its work-force to the same levels as its international competitors is a matter of grave concern to both the Government and industry. This situation prevails despite the billions of pounds which have been invested in training programmes for both those in work and the unwaged. If a different approach to training is not adopted, this country will fail to compete effectively in the international market-place.

There is currently an unacceptably high level of unemployment in the UK and yet, at the same time, there are urgent skills shortages. A skills mismatch exists between those who are unwaged and available for employment and the vacant jobs. In spite of this, training programmes, on the whole, are still remedial and short-term in their approach, and there seems to be no long-term strategic plan to improve this situation.

In many other countries, the people and their governments see education and training as a gateway to independence, success and opportunity. In this country, there is, generally speaking, a 'poverty of aspiration' which means that many people over 16 fail to invest in their own further learning. The existing education and training systems have failed to convince them that education, as well as being beneficial to the development of the individual learner, can be a doorway to prosperity.

The world of work is changing. People can no longer expect to have the same job for the whole of their lives. Many want to retire

PART IV: SUMMARY AND RECOMMENDATIONS

early, have more leisure time or work part-time. Technology is advancing; new skills are emerging and a flexible, competent, well-trained work-force is needed to cope and compete. The demographic trend means that there will be fewer young people available for work and this will also require a different approach to work and training.

A European dimension is about to be added to vocational education and training. After 1992, people and industry will be competing with other European countries, many of which have a much higher investment in education and training than this country.

All of these points make it clear that the UK's future economic success will be seriously impaired by continuing to have an underfunded, narrowly focused system of vocational education and training which only trains specific people for specific jobs. At the moment, people are not being prepared for the identified needs of the future, current training schemes are not eliminating unemployment and reducing skills shortages, there is not a high enough level of investment in learning and training and, most importantly, there is no learning culture. Short-term remedies do not recognize the long-term potential of people, nor do they build for a healthy economy in the future. Above all, central Government policy and practice needs to recognize this and take positive action to replace some of the more hollow rhetoric and limited resourcing currently allocated to this whole area.

So, what needs to be done? What kind of ideas will inform and shape future practice? How can plans and initiatives be implemented to the benefit of both the unwaged learner and the economy?

Shaping the future

There is no doubt that the message that 'training pays' is gaining momentum despite, or perhaps because of, the current difficult economic situation. Many organizations, employers, education and training institutions and TECs are now encouraging investment in people.

Numerous measures have been introduced to encourage employers in particular to improve and increase their commitment to training and to recognize people as the key resource in generating economic growth and development.

Underlying this recognition is the need to satisfy economic demands. This book, however, has stressed the shortcomings of a hard, purely economic approach, emphasizing instead the importance of a duality of purpose in the education and training of unwaged adults. This means placing the needs of people on an equal footing with the needs of the economy.

What has been apparent from the issues addressed in this book and reflected in the case-studies is that:

- the development of a learning culture in the UK, underpinned by issues of entitlement and rights of access by all adults, is the only way people will realize their full potential and, thereby, fully contribute to the growth and development of the economy;
- all unwaged adults need to have the same basic rights to quality vocational education and training as those who are employed;
- all unwaged adults should be given a minimum entitlement to this vocational education and training;
- all unwaged adults should be given the necessary impartial advice and guidance to enable them to make informed choices;
- there should be a recognition of the individuality and life experiences of unwaged adults, so that their full potential is realized;
- clear access routes should be created into vocational education and training so that unwaged adults can develop and acquire new skills throughout their lives;
- financial support should be available for all adults to ensure equity of opportunity and success; and
- sufficient and appropriate funding is essential to support quality vocational education and training for all unwaged adults.

If a better skilled and educated work-force is really needed, then the above mentioned goals are indispensable. To achieve them calls for a variety of key actions by individuals, organizations and the Government. Only a shared commitment and collaboration will bring about the changes advocated in this book. The measures necessary are:

- the introduction of legislation which will give adults the necessary entitlement and rights to education and training;
- the promotion of the idea of lifelong learning by those involved in the development of initial, compulsory schooling by helping to develop programmes, skills and qualifications which will ultimately give people greater control and choice in their personal and working lives;
- a real financial commitment to vocational education and training to bring this country into line with other nations;
- the creation of a compulsory, well-financed and supported staff development programme for all those working with unwaged

adults so that they understand the issues and can be instrumental in the developments described;
- the creation of media initiatives which reinforce the idea of lifelong learning and make unwaged adults, in particular, aware of the opportunities that are available to them;
- a co-ordinated approach to vocational education and training by colleges, TECs, training organizations and the voluntary sector needs to be established so that clear progression routes can be identified, opportunities can be effectively marketed and targeted, and resources used to the best advantage;
- the creation of a long-term strategy for the implementation of vocational education and training so that ideas about the development of the individual and lifelong learning can be developed to replace the existing short-term, unsatisfactory programmes; and
- the creation of 'adult friendly' environments, courses and programmes of work.

These measures will ultimately enhance individual skills which will, in turn, enhance economic performance. If they are implemented successfully, they will give unwaged adults greater flexibility, influence and power in their role in the labour market than before.

It is hoped that the ideas, information, examples of good practice and proposals put forward in this book will go some way towards creating the changes needed in vocational education and training for unwaged adults and stimulate new attitudes and approaches.

Bibliography

Baker, Kenneth (1989) Speech to NATFHE.
Ball, Sir Christopher (1990) *More Means Different*, RSA.
CBI Vocational Education and Training Task Force (1989) *Towards a Skills Revolution*.
Cassels, John (1990) *Britain's Real Skills Shortages and what to do about them*, Policy Studies Institute.
CAITS (1989) *Scrounging for Skills – Trainees' Views on ET in London*, Centre for Alternative Industrial and Technological Systems.
Charnley, A. H., McGivney, V. K. and Sims, D. J. (1985) *Education for the Adult Unemployed: Some Responses*, The National Institute of Adult Continuing Education.
City & Guilds: A comparison of NVQs and traditional qualifications – taken from a collection of information sheets compiled by City & Guilds.
Council of Europe (1990) *Report of the Topic Group on the Long-term Unemployed*, Council of Europe document DECS/EES (90) 11, prepared by Gerald Bogard.
Crossland, Derek (1991) *The Assessment of Prior Learning and Achievement*, REPLAN.
European Commission (1989) *Community Charter of Fundamental Social Rights of Workers*, adopted at Strasbourg, December 1989.
European Commission (1990) *Official Journal of the European Communities*, No. I 156/1.
FEU (1987) *Supporting Adult Learning – a discussion document*, Further Education Unit.
FEU (1990) *Curriculum Development through YTS Modular Credit Accumulation*, Further Education Unit.
Fiddy, Rob (ed.) (1983) *In Place of Work – Policy and Provision for the Young Unemployed*, Lewes: The Falmer Press.
Funnel, Peter and Müller, Dave (1991) *Vocational Education and the Challenge*, London: Kogan Page.
Gleeson, Denis (1990) *Training and its Alternatives*, Milton Keynes: Open University Press.
Hayes, J. and Nutman, P. (1981) *Understanding the Unemployed*, Tavistock.
Heinemann, K. (1978) *Arbeitslose Jugendliche*, Darmstadt: Luchterland.
HMSO (1988) *Employment for the 1990s*, White Paper.
HMSO (1991) *Education and Training for the 21st Century*, White Paper.
Institute of Employment Research (1989) University of Warwick.
Jackson, Michael (1985) *Youth Unemployment*, Croom Helm.

BIBLIOGRAPHY

Kanter, Dr Rosabeth Moss (1990) Taken from *The Prospect*, quarterly newsletter from the Prospect Centre, Issue No 20, July.

Killeen, John, White, Michael and Watts, A. G. (1991) *The Economic Value of Careers Guidance*, Policy Studies Unit in association with the National Institute for Careers Education and Counselling.

Lengrand, P. (nd) *Lifelong Education: Growth of the Concept*.

Levy, Margaret (1982–85) *The Core Skills Project and Work Based Learning*.

Miliband, David (1990) *Learning by Right – An Entitlement to Paid Education and Training*, Institute for Public Policy Research.

NALGO (1990) *NVQs are coming*, NALGO Education.

NCVQ: fundamental criteria for NVQs, taken from NCVQ guidelines.

NIACE (1990) *The Case for Adult Learners – Education and Training for the 21st Century*, The National Institute of Adult Continuing Education.

Read, Mel and Simpson, Alan (1991) *Against a Rising Tide – Racism, Europe and 1992*, Spokesman Press for Nottingham Racial Equality Council and the European Labour Forum.

REPLAN (1988a) *Second Chance to Learn? A review of WEA 'Second Chance to Learn' courses*, FEU-REPLAN.

REPLAN (1988b) *Working Through Words*, NIACE-REPLAN.

Scarman, The Rt. Hon. Lord (1981) *The Brixton Disorders*, HMSO.

Small, Nick (1984) *The Adult Training Strategy: Master Plan or Minor Expedient*, Association for Recurrent Education.

Thurley, Prof Keith (1990) Taken from *The Prospect*, quarterly newsletter from the Prospect Centre, Issue No. 20, July.

Tuckett, Alan (1991) *Towards a Learning Workforce: A policy discussion paper on adult learners at work*, NIACE.

UDACE (1986) *The Challenge of Change, Developing Educational Guidance for Adults*, Unit for the Development of Adult and Continuing Education.

Warr, P. (1983) 'Work, Jobs and Unemployment', *Bulletin of the British Psychological Society*, Volume 36, September 1983.

Watts, Alan and Knasel, Eddie (1985) *Adult Unemployment and the Curriculum – A Manual for Practitioners*, FEU/REPLAN.

Wellington, (1987) 'Stretching the Point' *Times Educational Supplement*, December.